I0474070

Albert Einstein at the age of three (1882)

Nikola Tesla.

The New Physics & Special and General Relativity:
An Original Compilation

BY
LUCIEN POINCARE
&
ALBERT EINSTEIN

©2012

This book consists of the following Two Titles:

THE NEW PHYSICS
AND ITS EVOLUTION
BY
LUCIEN POINCARÉ
Inspéctéur-General de l'Instruction Publique
Being the Authorized Translation of
"LA PHYSIQUE MODERNE, SON ÉVOLUTION"
NEW YORK
D. APPLETON AND COMPANY
©1909

RELATIVITY:
THE SPECIAL AND GENERAL THEORY

BY
ALBERT EINSTEIN
Translated By Robert W. Lawson

Publisher: Methuen & Co Ltd
First Published: December, 1916
Revised Edition
©1916.
Written: 1916
(This Revised Edition: 1924, 2012)
First Published: December, 1916
Translated: Robert W. Lawson (Authorized Translation)

THE NEW PHYSICS
AND ITS EVOLUTION
BY
LUCIEN POINCARÉ
Inspéctéur-General de l'Instruction Publique
Being the Authorized Translation of
"LA PHYSIQUE MODERNE, SON ÉVOLUTION"
NEW YORK
D. APPLETON AND COMPANY
©1909

Table of Contents

Prefatory Note

M. Lucien Poincaré is one of the distinguished family of mathematicians which has during the last few years given a Minister of Finance to the Republic and a President to the Académie des Sciences. He is also one of the nineteen Inspectors-General of Public Instruction who are charged with the duty of visiting the different universities and *lycées* in France and of reporting upon the state of the studies there pursued. Hence he is in an excellent position to appreciate at its proper value the extraordinary change which has lately revolutionized physical science, while his official position has kept him aloof from the controversies aroused by the discovery of radium and by recent speculations on the constitution of matter.

M. Poincaré's object and method in writing the book are sufficiently explained in the preface which follows; but it may be remarked that the best of methods has its defects, and the excessive condensation which has alone made it possible to include the last decade's discoveries in physical science within a compass of some 300 pages has, perhaps, made the facts here noted assimilable with difficulty by the untrained reader. To remedy this as far as possible, I have prefixed to the present translation a table of contents so extended as to form a fairly complete digest of the book, while full indexes of authors and subjects have also been added. The few notes necessary either for better elucidation of the terms employed, or for giving account of discoveries made while these pages were passing through the press, may be distinguished from the author's own by the signature "ED."

THE EDITOR.
ROYAL INSTITUTION OF GREAT BRITAIN, April 1907.

Author's Preface

During the last ten years so many works have accumulated in the domain of Physics, and so many new theories have been propounded, that those who follow with interest the progress of science, and even some professed scholars, absorbed as they are in their own special studies, find themselves at sea in a confusion more apparent than real.

It has therefore occurred to me that it might be useful to write a book which, while avoiding too great insistence on purely technical details, should try to make known the general results at which physicists have lately arrived, and to indicate the direction and import which should be ascribed to those speculations on the constitution of matter, and the discussions on the nature of first principles, to which it has become, so to speak, the fashion of the present day to devote oneself.

I have endeavoured throughout to rely only on the experiments in which we can place the most confidence, and, above all, to show how the ideas prevailing at the present day have been formed, by tracing their evolution, and rapidly examining the successive transformations which have brought them to their present condition.

In order to understand the text, the reader will have no need to consult any treatise on physics, for I have throughout given the necessary definitions and set forth the fundamental facts. Moreover, while strictly employing exact expressions, I have avoided the use of mathematical language. Algebra is an admirable tongue, but there are many occasions where it can only be used with much discretion.

Nothing would be easier than to point out many great omissions from this little volume; but some, at all events, are not involuntary.

Certain questions which are still too confused have been put on one side, as have a few others which form an important collection for a special study to be possibly made later. Thus, as regards electrical phenomena, the relations between electricity and optics, as also the theories of ionization, the electronic hypothesis, etc., have been treated at some length; but it has not been thought necessary to dilate upon the modes of production and utilization of the current, upon the phenomena of magnetism, or upon all the applications which belong to the domain of Electrotechnics.

L. POINCARÉ.

CHAPTER I
THE EVOLUTION OF PHYSICS

Revolutionary change in modern Physics only apparent: evolution not revolution the rule in Physical Theory— Revival of metaphysical speculation and influence of Descartes: all phenomena reduced to matter and movement— Modern physicists challenge this: physical, unlike mechanical, phenomena seldom reversible—Two schools, one considering experimental laws imperative, the other merely studying relations of magnitudes: both teach something of truth—Third or eclectic school— Is mechanics a branch of electrical science?

CHAPTER II
MEASUREMENTS

§ 1. *Metrology*: Lord Kelvin's view of its necessity— Its definition § 2. *The Measure of Length*: Necessity for unit— Absolute length—History of Standard—Description of Standard Metre—Unit of wave-lengths preferable—The International Metre § 3. *The Measure of Mass*: Distinction between mass and weight—Objections to legal kilogramme and its precision—Possible improvement § 4. *The Measure of Time*: Unit of time the second—Alternative units proposed—Improvements in chronometry and invar § 5. *The Measure of Temperature:* Fundamental and derived units— Ordinary unit of temperature purely arbitrary—Absolute unit mass of H at pressure of 1 m. of Hg at 0° C.—Divergence of thermometric and thermodynamic scales—Helium thermometer for low, thermo-electric couple for high, temperatures—Lummer and Pringsheim's improvements in thermometry. § 6. *Derived Units and Measure of Energy:* Importance of erg as unit—Calorimeter usual means of determination—Photometric units. § 7. *Measure of Physical Constants:* Constant of gravitation—Discoveries of Cavendish, Vernon Boys, Eötvös, Richarz and Krigar-Menzel—Michelson's improvements on Fizeau and Foucault's experiments— Measure of speed of light.

CHAPTER III
PRINCIPLES

§ 1. *The Principles of Physics:* The Principles of Mechanics affected by recent discoveries—Is mass indestructible?—Landolt and Heydweiller's experiments —Lavoisier's law only approximately true—Curie's principle of symmetry. § 2. *The Principle of the Conservation of Energy:* Its evolution: Bernoulli, Lavoisier and Laplace, Young, Rumford, Davy, Sadi Carnot, and Robert Mayer—Mayer's drawbacks—Error of those who would make mechanics part of energetics—Verdet's predictions—Rankine inventor of energetics—Usefulness of Work as standard form of energy—Physicists who think matter form of energy— Objections to this—Philosophical value of conservation doctrine. § 3. *The Principle of Carnot and Clausius:* Originality of Carnot's principle that fall of temperature necessary for production of work by heat— Clausius' postulate that heat cannot pass from cold to hot body without accessory phenomena—Entropy result of this—Definition of entropy—Entropy tends to increase incessantly—A magnitude which measures evolution of system—Clausius' and Kelvin's deduction that heat end of all energy in Universe—Objection to this— Carnot's principle not necessarily referable to mechanics —Brownian movements—Lippmann's objection to kinetic hypothesis. § 4. *Thermodynamics:* Historical work of Massieu, Willard Gibbs, Helmholtz, and Duhem—Willard Gibbs founder of thermodynamic statics, Van t'Hoff its reviver—The Phase Law—Raveau explains it without thermodynamics. § 5. *Atomism:* Connection of subject with preceding Hannequin's essay on the atomic hypothesis—Molecular physics in disfavour—Surface-tension, etc., vanishes when molecule reached—Size of molecule—Kinetic theory of gases—Willard Gibbs and Boltzmann introduce into it law of probabilities—Mean free path of gaseous molecules—Application to optics—Final division of matter.

CHAPTER IV
THE VARIOUS STATES OF MATTER

§ 1. *The Statics of Fluids*: Researches of Andrews, Cailletet, and others on liquid and gaseous states— Amagat's experiments—Van der Waals' equation—Discovery of corresponding states—Amagat's superposed diagrams—Exceptions to law—Statics of mixed fluids— Kamerlingh Onnes' researches—Critical Constants— Characteristic equation of fluid not yet ascertainable. § 2. *The Liquefaction of Gases and Low Temperatures*: Linde's, Siemens', and Claude's methods of liquefying gases—Apparatus of Claude described—Dewar's experiments—Modification of electrical properties of matter by extreme cold: of magnetic and chemical— Vitality of bacteria unaltered—Ramsay's discovery of rare gases of atmosphere—Their distribution in nature—Liquid hydrogen—Helium. § 3. *Solids and Liquids*: Continuity of Solid and Liquid States—Viscosity common to both—Also Rigidity— Spring's analogies of solids and liquids—Crystallization — Lehmann's liquid crystals—Their existence doubted —Tamman's view of discontinuity between crystalline and liquid states. § 4. *The Deformation of Solids*: Elasticity— Hoocke's, Bach's, and Bouasse's researches—Voigt on the elasticity of crystals—Elastic and permanent deformations—Brillouin's states of unstable equilibria—Duhem and the thermodynamic postulates—Experimental confirmation—Guillaume's researches on nickel steel—Alloys.

CHAPTER V
SOLUTIONS AND ELECTROLYTIC DISSOCIATION

§ 1. *Solution*: Kirchhoff's, Gibb's, Duhem's and Van t'Hoff's researches. § 2. *Osmosis*: History of phenomenon—Traube and biologists establish existence of semi-permeable walls—Villard's experiments with gases—Pfeffer shows osmotic pressure proportional to concentration—Disagreement as to cause of phenomenon. § 3. *Osmosis applied to Solution*: Van t'Hoff's discoveries—Analogy between dissolved body and perfect gas—Faults in analogy. § 4. *Electrolytic Dissociation*: Van t'Hoff's and Arrhenius' researches—Ionic hypothesis of—Fierce opposition to at first—Arrhenius' ideas now triumphant —Advantages of Arrhenius' hypothesis—"The ions which react"—Ostwald's conclusions from this—Nernst's theory of Electrolysis—Electrolysis of gases makes electronic theory probable—Faraday's two laws—Valency— Helmholtz's consequences from Faraday's laws.

CHAPTER VI
THE ETHER

§ 1. *The Luminiferous Ether*: First idea of Ether due to Descartes—Ether must be imponderable—Fresnel shows light vibrations to be transverse—Transverse vibrations cannot exist in fluid—Ether must be discontinuous. § 2. *Radiations*: Wave-lengths and their measurements—Rubens' and Lenard's researches— Stationary waves and colour-photography—Fresnel's hypothesis opposed by Neumann—Wiener's and Cotton's experiments. § 3. *TheElectromagnetic Ether*: Ampère's advocacy of mathematical expression—Faraday first shows influence of medium in electricity—Maxwell's proof that light-waves electromagnetic—His unintelligibility—Required confirmation of theory by Hertz. § 4. *Electrical Oscillations*: Hertz's experiments— Blondlot proves electromagnetic disturbance propagated with speed of light—Discovery of ether waves intermediate between Hertzian and visible ones—Rubens' and Nichols' experiments—Hertzian and light rays contrasted—Pressure of light. § 5. *The X-Rays*: Röntgen's discovery—Properties of X-rays—Not homogeneous—Rutherford and M'Clung's experiments on energy corresponding to—Barkla's experiments on polarisation of—Their speed that of light—Are they merely

Negative ions 1/1000 of size of atoms—Natural unit of electricity or electrons. § 3. *How Ions are Produced:* Various causes of ionization—Moreau's experiments with alkaline salts—Barus and Bloch on ionization by phosphorus vapours—Ionization always result of shock. § 4. *Electrons in Metals:* Movement of electrons in metals foreshadowed by Weber—Giese's, Riecke's, Drude's, and J.J. Thomson's researches—Path of ions in metals and conduction of heat—Theory of Lorentz—Hesehus' explanation of electrification by contact—Emission of electrons by charged body—Thomson's measurement of positive ions.

CHAPTER IX
CATHODE RAYS AND RADIOACTIVE BODIES

§ 1. *The Cathode Rays:* History of discovery—Crookes' theory—Lenard rays—Perrin's proof of negative charge—Cathode rays give rise to X-rays—The canal rays—Villard's researches and magneto-cathode rays—Ionoplasty—Thomson's measurements of speed of rays —All atoms can be dissociated. § 2. *Radioactive Substances:* Uranic rays of Niepce de St Victor and Becquerel—General radioactivity of matter—Le Bon's and Rutherford's comparison of uranic with X rays—Pierre and Mme. Curie's discovery of polonium and radium—Their characteristics—Debierne discovers actinium. § 3. *Radiations and Emanations of Radioactive Bodies:* Giesel's, Becquerel's, and Rutherford's Researches—Alpha, beta, and gamma rays—Sagnac's secondary rays—Crookes' spinthariscope—The emanation —Ramsay and Soddy's researches upon it—Transformations of radioactive bodies—Their order. § 4. *Disaggregation of Matter and Atomic Energy:* Actual transformations of matter in radioactive bodies —Helium or lead final product—Ultimate disappearance of radium from earth—Energy liberated by radium: its amount and source—Suggested models of radioactive atoms—Generalization from radioactive phenomena -Le Bon's theories—Ballistic hypothesis generally admitted—Does energy come from without—Sagnac's experiments—Elster and Geitel's *contra.*

CHAPTER X
THE ETHER AND MATTER

§ 1. *The Relations between the Ether and Matter:* Attempts to reduce all matter to forms of ether—Emission and absorption phenomena show reciprocal action— Laws of radiation—Radiation of gases—Production of spectrum—Differences between light and sound variations show difference of media—Cauchy's, Briot's, Carvallo's and Boussinesq's researches—Helmholtz's and Poincaré's electromagnetic theories of dispersion. § 2. *The*

CHAPTER XI
THE FUTURE OF PHYSICS

The New Physics and its Evolution

CHAPTER I
THE EVOLUTION OF PHYSICS

The now numerous public which tries with some success to keep abreast of the movement in science, from seeing its mental habits every day upset, and from occasionally witnessing unexpected discoveries that produce a more lively sensation from their reaction on social life, is led to suppose that we live in a really exceptional epoch, scored by profound crises and illustrated by extraordinary discoveries, whose singularity surpasses everything known in the past. Thus we often hear it said that physics, in particular, has of late years undergone a veritable revolution; that all its principles have been made new, that all the edifices constructed by our fathers have been overthrown, and that on the field thus cleared has sprung up the most abundant harvest that has ever enriched the domain of science.

It is in fact true that the crop becomes richer and more fruitful, thanks to the development of our laboratories, and that the quantity of seekers has considerably increased in all countries, while their quality has not diminished. We should be sustaining an absolute paradox, and at the same time committing a crying injustice, were we to contest the high importance of recent progress, and to seek to diminish the glory of contemporary physicists. Yet it may be as well not to give way to exaggerations, however pardonable, and to guard against facile illusions. On closer examination it will be seen that our predecessors might at several periods in history have conceived, as legitimately as ourselves, similar sentiments of scientific pride, and have felt that the world was about to appear to them transformed and under an aspect until then absolutely unknown.

Let us take an example which is salient enough; for, however arbitrary the conventional division of time may appear to a physicist's eyes, it is natural, when instituting a comparison between two epochs, to choose those which extend over a space of half a score of years, and are separated from each other by the gap of a century. Let us, then, go back a hundred years and examine what would have been the state of mind of an erudite amateur who had read and understood the chief publications on physical research between 1800 and 1810.

Let us suppose that this intelligent and attentive spectator witnessed in 1800 the discovery of the galvanic battery by Volta. He might from that moment have felt a presentiment that a prodigious transformation was about to occur in our mode of regarding electrical phenomena. Brought up in the ideas of Coulomb and Franklin, he might till then have imagined that electricity had unveiled nearly all its mysteries, when an entirely original apparatus suddenly gave birth to applications of the highest interest, and excited the blossoming of theories of immense philosophical extent.

In the treatises on physics published a little later, we find traces of the astonishment produced by this sudden revelation of a new world. "Electricity," wrote the Abbé Haüy, "enriched by the labour of so many distinguished physicists, seemed to have reached the term when a science has no further important steps before it, and only leaves to those who cultivate it the hope of confirming the discoveries of their predecessors, and of casting a brighter light on the truths revealed. One would have thought that all researches for diversifying the results of experiment were exhausted, and that theory itself could only be augmented by the addition of a greater degree of precision to the applications of principles already known. While science thus appeared to be making for repose, the phenomena of the convulsive movements observed by Galvani in the muscles of a frog when connected by metal were brought to the attention and astonishment of physicists.... Volta, in that Italy which had been the cradle of the new knowledge, discovered the principle of its true theory in a fact which reduces the explanation of all the phenomena in question to the simple contact of two substances of different nature. This fact became in his hands the germ of the admirable apparatus to which its manner of being and its fecundity assign one of the chief places among those with which the genius of mankind has enriched physics."

Shortly afterwards, our amateur would learn that Carlisle and Nicholson had decomposed water by the aid of a battery; then, that Davy, in 1803, had produced, by the help of the same battery, a quite unexpected phenomenon, and had succeeded in preparing metals endowed with marvellous properties, beginning with substances of an earthy appearance which had been known for a long time, but whose real nature had not been discovered.

In another order of ideas, surprises as prodigious would wait for our amateur. Commencing with 1802, he might have read the admirable series of memoirs which Young then published, and might thereby have learned how the study of the phenomena of diffraction led to the belief that the undulation theory, which, since the works of Newton seemed irretrievably condemned, was, on the contrary, beginning quite a new life. A little later—in 1808—he might have witnessed the discovery made by Malus of polarization by reflexion, and would have been able to note, no doubt with stupefaction, that under certain conditions a ray of light loses the property of being reflected.

He might also have heard of one Rumford, who was then promulgating very singular ideas on the nature of heat, who thought that the then classical notions might be false, that caloric does not exist as a fluid, and who, in 1804, even demonstrated that heat is created by friction. A few years later he would learn that Charles had enunciated a capital law on the dilatation of gases; that Pierre Prevost, in 1809, was making a study, full of original ideas, on radiant heat. In the meantime he would not have failed to read volumes iii. and iv. of the *Mecanique celeste* of Laplace, published in 1804 and 1805, and he might, no doubt, have thought that before long mathematics would enable physical science to develop with unforeseen safety.

All these results may doubtless be compared in importance with the present discoveries. When strange metals like potassium and sodium were isolated by an entirely new method, the astonishment must have been on a par with that caused in our time by the magnificent discovery of radium. The polarization of light is a phenomenon as undoubtedly singular as the existence of the X rays; and the upheaval produced in natural philosophy by the theories of the disintegration of matter and the ideas concerning electrons is probably not more considerable than that produced in the theories of light and heat by the works of Young and Rumford.

If we now disentangle ourselves from contingencies, it will be understood that in reality physical science progresses by evolution rather than by revolution. Its march is continuous. The facts which our theories enable us to discover, subsist and are linked together long after these theories have disappeared. Out of the materials of former edifices overthrown, new dwellings are constantly being reconstructed.

The labour of our forerunners never wholly perishes. The ideas of yesterday prepare for those of to-morrow; they contain them, so to speak, *in potentia*. Science is in some sort a living organism, which gives birth to an indefinite series of new beings taking the places of the old, and which evolves according to the nature of its environment, adapting itself to external conditions, and healing at every step the wounds which contact with reality may have occasioned.

Sometimes this evolution is rapid, sometimes it is slow enough; but it obeys the ordinary laws. The wants imposed by its surroundings create certain organs in science. The problems set to physicists by the engineer who wishes to facilitate transport or to produce better illumination, or by the doctor who seeks to know how such and such a remedy acts, or, again, by the physiologist desirous of understanding the mechanism of the gaseous and liquid exchanges between the cell and the outer medium, cause new chapters in physics to appear, and suggest researches adapted to the necessities of actual life.

The evolution of the different parts of physics does not, however, take place with equal speed, because the circumstances in which they are placed are not equally favourable. Sometimes a whole series of questions will appear forgotten, and will live only with a languishing existence; and then some accidental circumstance suddenly brings them new life, and they become the object of manifold labours, engross public attention, and invade nearly the whole domain of science.

We have in our own day witnessed such a spectacle. The discovery of the X rays—a discovery which physicists no doubt consider as the logical outcome of researches long pursued by a few scholars working in silence and obscurity on an otherwise much neglected subject—seemed to the public eye to have inaugurated a new era in the history of physics. If, as is the case, however, the extraordinary scientific movement provoked by Röntgen's sensational experiments has a very remote origin, it has, at least, been singularly quickened by the favourable conditions created by the interest aroused in its astonishing applications to radiography.

A lucky chance has thus hastened an evolution already taking place, and theories previously outlined have received a singular development. Without wishing to yield too much to what may be considered a whim of fashion, we cannot, if we are to note in this book the stage actually reached in the continuous march of physics, refrain from giving a clearly preponderant place to the questions suggested by the study of the new radiations. At the present time it is these questions which move us the most; they have shown us unknown horizons, and towards the fields recently opened to scientific activity the daily increasing crowd of searchers rushes in rather disorderly fashion.

One of the most interesting consequences of the recent discoveries has been to rehabilitate in the eyes of scholars, speculations relating to the constitution of matter, and, in a more general way, metaphysical problems. Philosophy has, of course, never been completely separated from science; but in times past many physicists dissociated themselves from studies which they looked upon as unreal word-squabbles, and sometimes not unreasonably abstained from joining in discussions which seemed to them idle and of rather puerile subtlety. They had seen the ruin of most of the systems built up *a priori* by daring philosophers, and deemed it more prudent to listen to the advice given by Kirchhoff and "to substitute the description of facts for a sham explanation of nature."

It should however be remarked that these physicists somewhat deceived themselves as to the value of their caution, and that the mistrust they manifested towards philosophical speculations did not preclude their admitting, unknown to themselves, certain axioms which they did not discuss, but which are, properly speaking, metaphysical conceptions. They were unconsciously speaking a language taught them by their predecessors, of which they made no attempt to discover the origin. It is thus that it was readily considered evident that physics must necessarily some day re-enter the domain of mechanics, and thence it was postulated that everything in nature is due to movement. We, further, accepted the principles of the classical mechanics without discussing their legitimacy.

This state of mind was, even of late years, that of the most illustrious physicists. It is manifested, quite sincerely and without the slightest reserve, in all the classical works devoted to physics. Thus Verdet, an illustrious professor who has had the greatest and most happy influence on the intellectual formation of a whole generation of scholars, and whose works are even at the present day very often consulted, wrote: "The true problem of the physicist is always to reduce all phenomena to that which seems to us the simplest and clearest, that is to say, to movement." In his celebrated course of lectures at l'École Polytechnique, Jamin likewise said: "Physics will one day form a chapter of general mechanics;" and in the preface to his excellent course of lectures on physics, M. Violle, in 1884, thus expresses himself: "The science of nature tends towards mechanics by a necessary evolution, the physicist being able to establish solid theories only on the laws of movement." The same idea is again met with in the words of Cornu in 1896: "The general tendency should be to show how the facts observed and the phenomena measured, though first brought together by empirical laws, end, by the impulse of successive progressions, in coming under the general laws of rational mechanics;" and the same physicist showed clearly that in his mind this connexion of phenomena with mechanics had a deep and philosophical reason, when, in the fine discourse pronounced by him at the opening ceremony of the Congrès de Physique in 1900, he exclaimed: "The mind of Descartes soars over modern physics, or rather, I should say, he is their luminary. The further we penetrate into the knowledge of natural phenomena, the clearer and the more developed becomes the bold Cartesian conception regarding the mechanism of the universe. There is nothing in the physical world but matter and movement."

If we adopt this conception, we are led to construct mechanical representations of the material world, and to imagine movements in the different parts of bodies capable of reproducing all the manifestations of nature. The kinematic knowledge of these movements, that is to say, the determination of the position, speed, and acceleration at a given moment of all the parts of the system, or, on the other hand, their dynamical study, enabling us to know what is the action of these parts on each other, would then be sufficient to enable us to foretell all that can occur in the domain of nature.

This was the great thought clearly expressed by the Encyclopædists of the eighteenth century; and if the necessity of interpreting the phenomena of electricity or light led the physicists of last century to imagine particular fluids which seemed to obey with some difficulty the ordinary rules of mechanics, these physicists still continued to retain their hope in the future, and to treat the idea of Descartes as an ideal to be reached sooner or later.

Certain scholars—particularly those of the English School—outrunning experiment, and pushing things to extremes, took pleasure in proposing very curious mechanical models which were often strange images of reality. The most illustrious of them, Lord Kelvin, may be considered as their representative type, and he has himself said: "It seems to me that the true sense of the question, Do we or do we not understand a particular subject in physics? is—Can we make a mechanical model which corresponds to it? I am never satisfied so long as I have been unable to make a mechanical model of the object. If I am able to do so, I understand it. If I cannot make such a model, I do not understand it." But it must be acknowledged that some of the models thus devised have become excessively complicated, and this complication has for a long time discouraged all but very bold minds. In addition, when it became a question of penetrating into the mechanism of molecules, and we were no longer satisfied to look at matter as a mass, the mechanical solutions seemed undetermined and the stability of the edifices thus constructed was insufficiently demonstrated.

Returning then to our starting-point, many contemporary physicists wish to subject Descartes' idea to strict criticism. From the philosophical point of view, they first enquire whether it is really demonstrated that there exists nothing else in the knowable than matter and movement. They ask themselves whether it is not habit and tradition in particular which lead us to ascribe to mechanics the origin of phenomena. Perhaps also a question of sense here comes in. Our senses, which are, after all, the only windows open towards external reality, give us a view of one side of the world only; evidently we only know the universe by the relations which exist between it and our organisms, and these organisms are peculiarly sensitive to movement.

Nothing, however, proves that those acquisitions which are the most ancient in historical order ought, in the development of science, to remain the basis of our knowledge. Nor does any theory prove that our perceptions are an exact indication of reality. Many reasons, on the contrary, might be invoked which tend to compel us to see in nature phenomena which cannot be reduced to movement.

Mechanics as ordinarily understood is the study of reversible phenomena. If there be given to the parameter which represents time,[1] and which has assumed increasing values during the duration of the phenomena, decreasing values which make it go the opposite way, the whole system will again pass through exactly the same stages as before, and all the phenomena will unfold themselves in reversed order. In physics, the contrary rule appears very general, and reversibility generally does not exist. It is an ideal and limited case, which may be sometimes approached, but can never, strictly speaking, be met with in its entirety. No physical phenomenon ever recommences in an identical manner if its direction be altered. It is true that certain mathematicians warn us that a mechanics can be devised in which reversibility would no longer be the rule, but the bold attempts made in this direction are not wholly satisfactory.

On the other hand, it is established that if a mechanical explanation of a phenomenon can be given, we can find an infinity of others which likewise account for all the peculiarities revealed by experiment. But, as a matter of fact, no one has ever succeeded in giving an indisputable mechanical representation of the whole physical world. Even were we disposed to admit the strangest solutions of the problem; to consent, for example, to be satisfied with the hidden systems devised by Helmholtz, whereby we ought to divide variable things into two classes, some accessible, and the others now and for ever unknown, we should never manage to construct an edifice to contain all the known facts. Even the very comprehensive mechanics of a Hertz fails where the classical mechanics has not succeeded.

Deeming this check irremediable, many contemporary physicists give up attempts which they look upon as condemned beforehand, and adopt, to guide them in their researches, a method which at first sight appears much more modest, and also much more sure. They make up their minds not to see at once to the bottom of things; they no longer seek to suddenly strip the last veils from nature, and to divine her supreme secrets; but they work prudently and advance but slowly, while on the ground thus conquered foot by foot they endeavour to establish themselves firmly. They study the various magnitudes directly accessible to their observation without busying themselves as to their essence. They measure quantities of heat and of temperature, differences of potential, currents, and magnetic fields; and then, varying the conditions, apply the rules of experimental method, and discover between these magnitudes mutual relations, while they thus succeed in enunciating laws which translate and sum up their labours.

These empirical laws, however, themselves bring about by induction the promulgation of more general laws, which are termed principles. These principles are originally only the results of experiments, and experiment allows them besides to be checked, and their more or less high degree of generality to be verified. When they have been thus definitely established, they may serve as fresh starting-points, and, by deduction, lead to very varied discoveries.

The principles which govern physical science are few in number, and their very general form gives them a philosophical appearance, while we cannot long resist the temptation of regarding them as metaphysical dogmas. It thus happens that the least bold physicists, those who have wanted to show themselves the most reserved, are themselves led to forget the experimental character of the laws they have propounded, and to see in them imperious beings whose authority, placed above all verification, can no longer be discussed.

Others, on the contrary, carry prudence to the extent of timidity. They desire to grievously limit the field of scientific investigation, and they assign to science a too restricted domain. They content themselves with representing phenomena by equations, and think that they ought to submit to calculation magnitudes experimentally determined, without asking themselves whether these calculations retain a physical meaning. They are thus led to reconstruct a physics in which there again appears the idea of quality, understood, of course, not in the scholastic sense, since from this quality we can argue with some precision by representing it under numerical symbols, but still constituting an element of differentiation and of heterogeneity.

Notwithstanding the errors they may lead to if carried to excess, both these doctrines render, as a whole, most important service. It is no bad thing that these contradictory tendencies should subsist, for this variety in the conception of phenomena gives to actual science a character of intense life and of veritable youth, capable of impassioned efforts towards the truth. Spectators who see such moving and varied pictures passing before them, experience the feeling that there no longer exist systems fixed in an immobility which seems that of death. They feel that nothing is unchangeable; that ceaseless transformations are taking place before their eyes; and that this continuous evolution and perpetual change are the necessary conditions of progress.

A great number of seekers, moreover, show themselves on their own account perfectly eclectic. They adopt, according to their needs, such or such a manner of looking at nature, and do not hesitate to utilize very different images when they appear to them useful and convenient. And, without doubt, they are not wrong, since these images are only symbols convenient for language. They allow facts to be grouped and associated, but only present a fairly distant resemblance with the objective reality. Hence it is not forbidden to multiply and to modify them according to circumstances. The really essential thing is to have, as a guide through the unknown, a map which certainly does not claim to represent all the aspects of nature, but which, having been drawn up according to predetermined rules, allows us to follow an ascertained road in the eternal journey towards the truth.

Among the provisional theories which are thus willingly constructed by scholars on their journey, like edifices hastily run up to receive an unforeseen harvest, some still appear very bold and very singular. Abandoning the search after mechanical models for all electrical phenomena, certain physicists reverse, so to speak, the conditions of the problem, and ask themselves whether, instead of giving a mechanical interpretation to electricity, they may not, on the contrary, give an electrical interpretation to the phenomena of matter and motion, and thus merge mechanics itself in electricity. One thus sees dawning afresh the eternal hope of co-ordinating all natural phenomena in one grandiose and imposing synthesis. Whatever may be the fate reserved for such attempts, they deserve attention in the highest degree; and it is desirable to examine them carefully if we wish to have an exact idea of the tendencies of modern physics.

CHAPTER II
MEASUREMENTS

§ 1. METROLOGY

Not so very long ago, the scholar was often content with qualitative observations. Many phenomena were studied without much trouble being taken to obtain actual measurements. But it is now becoming more and more understood that to establish the relations which exist between physical magnitudes, and to represent the variations of these magnitudes by functions which allow us to use the power of mathematical analysis, it is most necessary to express each magnitude by a definite number.

Under these conditions alone can a magnitude be considered as effectively known. "I often say," Lord Kelvin has said, "that if you can measure that of which you are speaking and express it by a number you know something of your subject; but if you cannot measure it nor express it by a number, your knowledge is of a sorry kind and hardly satisfactory. It may be the beginning of the acquaintance, but you are hardly, in your thoughts, advanced towards science, whatever the subject may be."

It has now become possible to measure exactly the elements which enter into nearly all physical phenomena, and these measurements are taken with ever increasing precision. Every time a chapter in science progresses, science shows itself more exacting; it perfects its means of investigation, it demands more and more exactitude, and one of the most striking features of modern physics is this constant care for strictness and clearness in experimentation.

A veritable science of measurement has thus been constituted which extends over all parts of the domain of physics. This science has its rules and its methods; it points out the best processes of calculation, and teaches the method of correctly estimating errors and taking account of them. It has perfected the processes of experiment, co-ordinated a large number of results, and made possible the unification of standards. It is thanks to it that the system of measurements unanimously adopted by physicists has been formed.

At the present day we designate more peculiarly by the name of metrology that part of the science of measurements which devotes itself specially to the determining of the prototypes representing the fundamental units of dimension and mass, and of the standards of the first order which are derived from them. If all measurable quantities, as was long thought possible, could be reduced to the magnitudes of mechanics, metrology would thus be occupied with the essential elements entering into all phenomena, and might legitimately claim the highest rank in science. But even when we suppose that some magnitudes can never be connected with mass, length, and time, it still holds a preponderating place, and its progress finds an echo throughout the whole domain of the natural sciences. It is therefore well, in order to give an account of the general progress of physics, to examine at the outset the improvements which have been effected in these fundamental measurements, and to see what precision these improvements have allowed us to attain.

§ 2. THE MEASURE OF LENGTH

To measure a length is to compare it with another length taken as unity. Measurement is therefore a relative operation, and can only enable us to know ratios. Did both the length to be measured and the unit chosen happen to vary simultaneously and in the same degree, we should perceive no change. Moreover, the unit being, by definition, the term of comparison, and not being itself comparable with anything, we have theoretically no means of ascertaining whether its length varies.

If, however, we were to note that, suddenly and in the same proportions, the distance between two points on this earth had increased, that all the planets had moved further from each other, that all objects around us had become larger, that we ourselves had become taller, and that the distance travelled by light in the duration of a vibration had become greater, we should not hesitate to think ourselves the victims of an illusion, that in reality all these distances had remained fixed, and that all these appearances were due to a shortening of the rule which we had used as the standard for measuring the lengths.

From the mathematical point of view, it may be considered that the two hypotheses are equivalent; all has lengthened around us, or else our standard has become less. But it is no simple question of convenience and simplicity which leads us to reject the one supposition and to accept the other; it is right in this case to listen to the voice of common sense, and those physicists who have an instinctive trust in the notion of an absolute length are perhaps not wrong. It is only by choosing our unit from those which at all times have seemed to all men the most invariable, that we are able in our experiments to note that the same causes acting under identical conditions always produce the same effects. The idea of absolute length is derived from the principle of causality; and our choice is forced upon us by the necessity of obeying this principle, which we cannot reject without declaring by that very act all science to be impossible.

Similar remarks might be made with regard to the notions of absolute time and absolute movement. They have been put in evidence and set forth very forcibly by a learned and profound mathematician, M. Painlevé.

On the particularly clear example of the measure of length, it is interesting to follow the evolution of the methods employed, and to run through the history of the progress in precision from the time that we have possessed authentic documents relating to this question. This history has been written in a masterly way by one of the physicists who have in our days done the most by their personal labours to add to it glorious pages. M. Benoit, the learned Director of the International Bureau of Weights and Measures, has furnished in various reports very complete details on the subject, from which I here borrow the most interesting.

We know that in France the fundamental standard for measures of length was for a long time the *Toise du Châtelet*, a kind of callipers formed of a bar of iron which in 1668 was embedded in the outside wall of the Châtelet, at the foot of the staircase. This bar had at its extremities two projections with square faces, and all the *toises* of commerce had to fit exactly between them. Such a standard, roughly constructed, and exposed to all the injuries of weather and time, offered very slight guarantees either as to the permanence or the correctness of its copies. Nothing, perhaps, can better convey an idea of the importance of the modifications made in the methods of experimental physics than the easy comparison between so rudimentary a process and the actual measurements effected at the present time.

The *Toise du Châtelet*, notwithstanding its evident faults, was employed for nearly a hundred years; in 1766 it was replaced by the *Toise du Pérou*, so called because it had served for the measurements of the terrestrial arc effected in Peru from 1735 to 1739 by Bouguer, La Condamine, and Godin. At that time, according to the comparisons made between this new *toise* and the *Toise du Nord*, which had also been used for the measurement of an arc of the meridian, an error of the tenth part of a millimetre in measuring lengths of the order of a metre was considered quite unimportant. At the end of the eighteenth century, Delambre, in his work *Sur la Base du Système métrique décimal*, clearly gives us to understand that magnitudes of the order of the hundredth of a millimetre appear to him incapable of observation, even in scientific researches of the highest precision. At the present date the International Bureau of Weights and Measures guarantees, in the determination of a standard of length compared with the metre, an approximation of two or three ten-thousandths of a millimetre, and even a little more under certain circumstances.

This very remarkable progress is due to the improvements in the method of comparison on the one hand, and in the manufacture of the standard on the other. M. Benoit rightly points out that a kind of competition has been set up between the standard destined to represent the unit with its subdivisions and multiples and the instrument charged with observing it, comparable, up to a certain point, with that which in another order of ideas goes on between the gun and the armour-plate.

The measuring instrument of to-day is an instrument of comparison constructed with meticulous care, which enables us to do away with causes of error formerly ignored, to eliminate the action of external phenomena, and to withdraw the experiment from the influence of even the personality of the observer. This standard is no longer, as formerly, a flat rule, weak and fragile, but a rigid bar, incapable of deformation, in which the material is utilised in the best conditions of resistance. For a standard with ends has been substituted a standard with marks, which permits much more precise definition and can be employed in optical processes of observation alone; that is, in processes which can produce in it no deformation and no alteration. Moreover, the marks are traced on the plane of the neutral fibres[2] exposed, and the invariability of their distance apart is thus assured, even when a change is made in the way the rule is supported.

Thanks to studies thus systematically pursued, we have succeeded in the course of a hundred years in increasing the precision of measures in the proportion of a thousand to one, and we may ask ourselves whether such an increase will continue in the future. No doubt progress will not be stayed; but if we keep to the definition of length by a material standard, it would seem that its precision cannot be considerably increased. We have nearly reached the limit imposed by the necessity of making strokes of such a thickness as to be observable under the microscope.

It may happen, however, that we shall be brought one of these days to a new conception of the measure of length, and that very different processes of determination will be thought of. If we took as unit, for instance, the distance covered by a given radiation during a vibration, the optical processes would at once admit of much greater precision.

Thus Fizeau, the first to have this idea, says: "A ray of light, with its series of undulations of extreme tenuity but perfect regularity, may be considered as a micrometer of the greatest perfection, and particularly suitable for determining length." But in the present state of things, since the legal and customary definition of the unit remains a material standard, it is not enough to measure length in terms of wave-lengths, and we must also know the value of these wave-lengths in terms of the standard prototype of the metre.

This was determined in 1894 by M. Michelson and M. Benoit in an experiment which will remain classic. The two physicists measured a standard length of about ten centimetres, first in terms of the wave-lengths of the red, green, and blue radiations of cadmium, and then in terms of the standard metre. The great difficulty of the experiment proceeds from the vast difference which exists between the lengths to be compared, the wave-lengths barely amounting to half a micron; [3] the process employed consisted in noting, instead of this length, a length easily made about a thousand times greater, namely, the distance between the fringes of interference.

In all measurement, that is to say in every determination of the relation of a magnitude to the unit, there has to be determined on the one hand the whole, and on the other the fractional part of this ratio, and naturally the most delicate determination is generally that of this fractional part. In optical processes the difficulty is reversed. The fractional part is easily known, while it is the high figure of the number representing the whole which becomes a very serious obstacle. It is this obstacle which MM. Michelson and Benoit overcame with admirable ingenuity. By making use of a somewhat similar idea, M. Macé de Lépinay and MM. Perot and Fabry, have lately effected by optical methods, measurements of the greatest precision, and no doubt further progress may still be made. A day may perhaps come when a material standard will be given up, and it may perhaps even be recognised that such a standard in time changes its length by molecular strain, and by wear and tear: and it will be further noted that, in accordance with certain theories which will be noticed later on, it is not invariable when its orientation is changed.

For the moment, however, the need of any change in the definition of the unit is in no way felt; we must, on the contrary, hope that the use of the unit adopted by the physicists of the whole world will spread more and more. It is right to remark that a few errors still occur with regard to this unit, and that these errors have been facilitated by incoherent legislation. France herself, though she was the admirable initiator of the metrical system, has for too long allowed a very regrettable confusion to exist; and it cannot be noted without a certain sadness that it was not until the *11th July 1903* that a law was promulgated re-establishing the agreement between the legal and the scientific definition of the metre.

Perhaps it may not be useless to briefly indicate here the reasons of the disagreement which had taken place. Two definitions of the metre can be, and in fact were given. One had for its basis the dimensions of the earth, the other the length of the material standard. In the minds of the founders of the metrical system, the first of these was the true definition of the unit of length, the second merely a simple representation. It was admitted, however, that this representation had been constructed in a manner perfect enough for it to be nearly impossible to perceive any difference between the unit and its representation, and for the practical identity of the two definitions to be thus assured. The creators of the metrical system were persuaded that the measurements of the meridian effected in their day could never be surpassed in precision; and on the other hand, by borrowing from nature a definite basis, they thought to take from the definition of the unit some of its arbitrary character, and to ensure the means of again finding the same unit if by any accident the standard became altered. Their confidence in the value of the processes they had seen employed was exaggerated, and their mistrust of the future unjustified. This example shows how imprudent it is to endeavour to fix limits to progress. It is an error to think the march of science can be stayed; and in reality it is now known that the ten-millionth part of the quarter of the terrestrial meridian is longer than the metre by 0.187 millimetres. But contemporary physicists do not fall into the same error as their forerunners, and they regard the present result as merely provisional. They guess, in fact, that new improvements will be effected in the art of measurement; they know that geodesical processes, though much improved in our days, have still much to do to attain the precision displayed in the construction and determination of standards of the first order; and consequently they do not propose to keep the ancient definition, which would lead to having for unit a magnitude possessing the grave defect from a practical point of view of being constantly variable.

We may even consider that, looked at theoretically, its permanence would not be assured. Nothing, in fact, proves that sensible variations may not in time be produced in the value of an arc of the meridian, and serious difficulties may arise regarding the probable inequality of the various meridians.

For all these reasons, the idea of finding a natural unit has been gradually abandoned, and we have become resigned to accepting as a fundamental unit an arbitrary and conventional length having a material representation recognised by universal consent; and it was this unit which was consecrated by the following law of the 11th July 1903:—

"The standard prototype of the metrical system is the international metre, which has been sanctioned by the General Conference on Weights and Measures."

§ 3. THE MEASURE OF MASS

On the subject of measures of mass, similar remarks to those on measures of length might be made. The confusion here was perhaps still greater, because, to the uncertainty relating to the fixing of the unit, was added some indecision on the very nature of the magnitude defined. In law, as in ordinary practice, the notions of weight and of mass were not, in fact, separated with sufficient clearness.

They represent, however, two essentially different things. Mass is the characteristic of a quantity of matter; it depends neither on the geographical position one occupies nor on the altitude to which one may rise; it remains invariable so long as nothing material is added or taken away. Weight is the action which gravity has upon the body under consideration; this action does not depend solely on the body, but on the earth as well; and when it is changed from one spot to another, the weight changes, because gravity varies with latitude and altitude.

These elementary notions, to-day understood even by young beginners, appear to have been for a long time indistinctly grasped. The distinction remained confused in many minds, because, for the most part, masses were comparatively estimated by the intermediary of weights. The estimations of weight made with the balance utilize the action of the weight on the beam, but in such conditions that the influence of the variations of gravity becomes eliminated. The two weights which are being compared may both of them change if the weighing is effected in different places, but they are attracted in the same proportion. If once equal, they remain equal even when in reality they may both have varied.

The current law defines the kilogramme as the standard of mass, and the law is certainly in conformity with the rather obscurely expressed intentions of the founders of the metrical system. Their terminology was vague, but they certainly had in view the supply of a standard for commercial transactions, and it is quite evident that in barter what is important to the buyer as well as to the seller is not the attraction the earth may exercise on the goods, but the quantity that may be supplied for a given price. Besides, the fact that the founders abstained from indicating any specified spot in the definition of the kilogramme, when they were perfectly acquainted with the considerable variations in the intensity of gravity, leaves no doubt as to their real desire.

The same objections have been made to the definition of the kilogramme, at first considered as the mass of a cubic decimetre of water at 4° C., as to the first definition of the metre. We must admire the incredible precision attained at the outset by the physicists who made the initial determinations, but we know at the present day that the kilogramme they constructed is slightly too heavy (by about 1/25,000). Very remarkable researches have been carried out with regard to this determination by the International Bureau, and by MM. Macé de Lépinay and Buisson. The law of the 11th July 1903 has definitely regularized the custom which physicists had adopted some years before; and the standard of mass, the legal prototype of the metrical system, is now the international kilogramme sanctioned by the Conference of Weights and Measures.

The comparison of a mass with the standard is effected with a precision to which no other measurement can attain. Metrology vouches for the hundredth of a milligramme in a kilogramme; that is to say, that it estimates the hundred-millionth part of the magnitude studied.

We may—as in the case of the lengths—ask ourselves whether this already admirable precision can be surpassed; and progress would seem likely to be slow, for difficulties singularly increase when we get to such small quantities. But it is permitted to hope that the physicists of the future will do still better than those of to-day; and perhaps we may catch a glimpse of the time when we shall begin to observe that the standard, which is constructed from a heavy metal, namely, iridium-platinum, itself obeys an apparently general law, and little by little loses some particles of its mass by emanation.

§ 4. THE MEASURE OF TIME

The third fundamental magnitude of mechanics is time. There is, so to speak, no physical phenomenon in which the notion of time linked to the sequence of our states of consciousness does not play a considerable part.

Ancestral habits and a very early tradition have led us to preserve, as the unit of time, a unit connected with the earth's movement; and the unit to-day adopted is, as we know, the sexagesimal second of mean time. This magnitude, thus defined by the conditions of a natural motion which may itself be modified, does not seem to offer all the guarantees desirable from the point of view of invariability. It is certain that all the friction exercised on the earth—by the tides, for instance—must slowly lengthen the duration of the day, and must influence the movement of the earth round the sun. Such influence is certainly very slight, but it nevertheless gives an unfortunately arbitrary character to the unit adopted.

We might have taken as the standard of time the duration of another natural phenomenon, which appears to be always reproduced under identical conditions; the duration, for instance, of a given luminous vibration. But the experimental difficulties of evaluation with such a unit of the times which ordinarily have to be considered, would be so great that such a reform in practice cannot be hoped for. It should, moreover, be remarked that the duration of a vibration may itself be influenced by external circumstances, among which are the variations of the magnetic field in which its source is placed. It could not, therefore, be strictly considered as independent of the earth; and the theoretical advantage which might be expected from this alteration would be somewhat illusory.

Perhaps in the future recourse may be had to very different phenomena. Thus Curie pointed out that if the air inside a glass tube has been rendered radioactive by a solution of radium, the tube may be sealed up, and it will then be noted that the radiation of its walls diminishes with time, in accordance with an exponential law. The constant of time derived by this phenomenon remains the same whatever the nature and dimensions of the walls of the tube or the temperature may be, and time might thus be denned independently of all the other units.

We might also, as M. Lippmann has suggested in an extremely ingenious way, decide to obtain measures of time which can be considered as absolute because they are determined by parameters of another nature than that of the magnitude to be measured. Such experiments are made possible by the phenomena of gravitation. We could employ, for instance, the pendulum by adopting, as the unit of force, the force which renders the constant of gravitation equal to unity. The unit of time thus defined would be independent of the unit of length, and would depend only on the substance which would give us the unit of mass under the unit of volume.

It would be equally possible to utilize electrical phenomena, and one might devise experiments perfectly easy of execution. Thus, by charging a condenser by means of a battery, and discharging it a given number of times in a given interval of time, so that the effect of the current of discharge should be the same as the effect of the output of the battery through a given resistance, we could estimate, by the measurement of the electrical magnitudes, the duration of the interval noted. A system of this kind must not be looked upon as a simple *jeu d'esprit*, since this very practicable experiment would easily permit us to check, with a precision which could be carried very far, the constancy of an interval of time.

From the practical point of view, chronometry has made in these last few years very sensible progress. The errors in the movements of chronometers are corrected in a much more systematic way than formerly, and certain inventions have enabled important improvements to be effected in the construction of these instruments. Thus the curious properties which steel combined with nickel—so admirably studied by M.Ch.Ed. Guillaume— exhibits in the matter of dilatation are now utilized so as to almost completely annihilate the influence of variations of temperature.

§ 5. THE MEASURE OF TEMPERATURE

From the three mechanical units we derive secondary units; as, for instance, the unit of work or mechanical energy. The kinetic theory takes temperature, as well as heat itself, to be a quantity of energy, and thus seems to connect this notion with the magnitudes of mechanics. But the legitimacy of this theory cannot be admitted, and the calorific movement should also be a phenomenon so strictly confined in space that our most delicate means of investigation would not enable us to perceive it. It is better, then, to continue to regard the unit of difference of temperature as a distinct unit, to be added to the fundamental units.

To define the measure of a certain temperature, we take, in practice, some arbitrary property of a body. The only necessary condition of this property is, that it should constantly vary in the same direction when the temperature rises, and that it should possess, at any temperature, a well-marked value. We measure this value by melting ice and by the vapour of boiling water under normal pressure, and the successive hundredths of its variation, beginning with the melting ice, defines the percentage. Thermodynamics, however, has made it plain that we can set up a thermometric scale without relying upon any determined property of a real body. Such a scale has an absolute value independently of the properties of matter. Now it happens that if we make use for the estimation of temperatures, of the phenomena of dilatation under a constant pressure, or of the increase of pressure in a constant volume of a gaseous body, we obtain a scale very near the absolute, which almost coincides with it when the gas possesses certain qualities which make it nearly what is called a perfect gas. This most lucky coincidence has decided the choice of the convention adopted by physicists. They define normal temperature by means of the variations of pressure in a mass of hydrogen beginning with the initial pressure of a metre of mercury at 0° C.

M.P. Chappuis, in some very precise experiments conducted with much method, has proved that at ordinary temperatures the indications of such a thermometer are so close to the degrees of the theoretical scale that it is almost impossible to ascertain the value of the divergences, or even the direction that they take. The divergence becomes, however, manifest when we work with extreme temperatures. It results from the useful researches of M. Daniel Berthelot that we must subtract +0.18° from the indications of the hydrogen thermometer towards the temperature -240° C, and add +0.05° to 1000° to equate them with the thermodynamic scale. Of course, the difference would also become still more noticeable on getting nearer to the absolute zero; for as hydrogen gets more and more cooled, it gradually exhibits in a lesser degree the characteristics of a perfect gas.

To study the lower regions which border on that kind of pole of cold towards which are straining the efforts of the many physicists who have of late years succeeded in getting a few degrees further forward, we may turn to a gas still more difficult to liquefy than hydrogen. Thus, thermometers have been made of helium; and from the temperature of -260° C. downward the divergence of such a thermometer from one of hydrogen is very marked.

The measurement of very high temperatures is not open to the same theoretical objections as that of very low temperatures; but, from a practical point of view, it is as difficult to effect with an ordinary gas thermometer. It becomes impossible to guarantee the reservoir remaining sufficiently impermeable, and all security disappears, notwithstanding the use of recipients very superior to those of former times, such as those lately devised by the physicists of the *Reichansalt*. This difficulty is obviated by using other methods, such as the employment of thermo-electric couples, such as the very convenient couple of M. le Chatelier; but the graduation of these instruments can only be effected at the cost of a rather bold extrapolation.

M.D. Berthelot has pointed out and experimented with a very interesting process, founded on the measurement by the phenomena of interference of the refractive index of a column of air subjected to the temperature it is desired to measure. It appears admissible that even at the highest temperatures the variation of the power of refraction is strictly proportional to that of the density, for this proportion is exactly verified so long as it is possible to check it precisely. We can thus, by a method which offers the great advantage of being independent of the power and dimension of the envelopes employed—since the length of the column of air considered alone enters into the calculation—obtain results equivalent to those given by the ordinary air thermometer.

Another method, very old in principle, has also lately acquired great importance. For a long time we sought to estimate the temperature of a body by studying its radiation, but we did not know any positive relation between this radiation and the temperature, and we had no good experimental method of estimation, but had recourse to purely empirical formulas and the use of apparatus of little precision. Now, however, many physicists, continuing the classic researches of Kirchhoff, Boltzmann, Professors Wien and Planck, and taking their starting-point from the laws of thermodynamics, have given formulas which establish the radiating power of a dark body as a function of the temperature and the wave-length, or, better still, of the total power as a function of the temperature and wave-length corresponding to the maximum value of the power of radiation. We see, therefore, the possibility of appealing for the measurement of temperature to a phenomenon which is no longer the variation of the elastic force of a gas, and yet is also connected with the principles of thermodynamics.

This is what Professors Lummer and Pringsheim have shown in a series of studies which may certainly be reckoned among the greatest experimental researches of the last few years. They have constructed a radiator closely resembling the theoretically integral radiator which a closed isothermal vessel would be, and with only a very small opening, which allows us to collect from outside the radiations which are in equilibrium with the interior. This vessel is formed of a hollow carbon cylinder, heated by a current of high intensity; the radiations are studied by means of a bolometer, the disposition of which varies with the nature of the experiments.

It is hardly possible to enter into the details of the method, but the result sufficiently indicates its importance. It is now possible, thanks to their researches, to estimate a temperature of 2000° C. to within about 5°. Ten years ago a similar approximation could hardly have been arrived at for a temperature of 1000° C.

§ 6. DERIVED UNITS AND THE MEASURE OF A QUANTITY OF ENERGY

It must be understood that it is only by arbitrary convention that a dependency is established between a derived unit and the fundamental units. The laws of numbers in physics are often only laws of proportion. We transform them into laws of equation, because we introduce numerical coefficients and choose the units on which they depend so as to simplify as much as possible the formulas most in use. A particular speed, for instance, is in reality nothing else but a speed, and it is only by the peculiar choice of unit that we can say that it is the space covered during the unit of time. In the same way, a quantity of electricity is a quantity of electricity; and there is nothing to prove that, in its essence, it is really reducible to a function of mass, of length, and of time.

Persons are still to be met with who seem to have some illusions on this point, and who see in the doctrine of the dimensions of the units a doctrine of general physics, while it is, to say truth, only a doctrine of metrology. The knowledge of dimensions is valuable, since it allows us, for instance, to easily verify the homogeneity of a formula, but it can in no way give us any information on the actual nature of the quantity measured.

Magnitudes to which we attribute like dimensions may be qualitatively irreducible one to the other. Thus the different forms of energy are measured by the same unit, and yet it seems that some of them, such as kinetic energy, really depend on time; while for others, such as potential energy, the dependency established by the system of measurement seems somewhat fictitious.

The numerical value of a quantity of energy of any nature should, in the system C.G.S., be expressed in terms of the unit called the erg; but, as a matter of fact, when we wish to compare and measure different quantities of energy of varying forms, such as electrical, chemical, and other quantities, etc., we nearly always employ a method by which all these energies are finally transformed and used to heat the water of a calorimeter. It is therefore very important to study well the calorific phenomenon chosen as the unit of heat, and to determine with precision its mechanical equivalent, that is to say, the number of ergs necessary to produce this unit. This is a number which, on the principle of equivalence, depends neither on the method employed, nor the time, nor any other external circumstance.

As the result of the brilliant researches of Rowland and of Mr Griffiths on the variations of the specific heat of water, physicists have decided to take as calorific standard the quantity of heat necessary to raise a gramme of water from 15° to 16° C., the temperature being measured by the scale of the hydrogen thermometer of the International Bureau.

On the other hand, new determinations of the mechanical equivalent, among which it is right to mention that of Mr. Ames, and a full discussion as to the best results, have led to the adoption of the number 4.187 to represent the number of ergs capable of producing the unit of heat.

In practice, the measurement of a quantity of heat is very often effected by means of the ice calorimeter, the use of which is particularly simple and convenient. There is, therefore, a very special interest in knowing exactly the melting-point of ice. M. Leduc, who for several years has measured a great number of physical constants with minute precautions and a remarkable sense of precision, concludes, after a close discussion of the various results obtained, that this heat is equal to 79.1 calories. An error of almost a calorie had been committed by several renowned experimenters, and it will be seen that in certain points the art of measurement may still be largely perfected.

To the unit of energy might be immediately attached other units. For instance, radiation being nothing but a flux of energy, we could, in order to establish photometric units, divide the normal spectrum into bands of a given width, and measure the power of each for the unit of radiating surface.

But, notwithstanding some recent researches on this question, we cannot yet consider the distribution of energy in the spectrum as perfectly known. If we adopt the excellent habit which exists in some researches of expressing radiating energy in ergs, it is still customary to bring the radiations to a standard giving, by its constitution alone, the unit of one particular radiation. In particular, the definitions are still adhered to which were adopted as the result of the researches of M. Violle on the radiation of fused platinum at the temperature of solidification; and most physicists utilize in the ordinary methods of photometry the clearly defined notions of M. Blondel as to the luminous intensity of flux, illumination (*éclairement*), light (*éclat*), and lighting (*éclairage*), with the corresponding units, decimal candle, *lumen*, *lux*, carcel lamp, candle per square centimetre, and *lumen*-hour. [4]

§ 7. MEASURE OF CERTAIN PHYSICAL CONSTANTS

The progress of metrology has led, as a consequence, to corresponding progress in nearly all physical measurements, and particularly in the measure of natural constants. Among these, the constant of gravitation occupies a position quite apart from the importance and simplicity of the physical law which defines it, as well as by its generality. Two material particles are mutually attracted to each other by a force directly proportional to the product of their mass, and inversely proportional to the square of the distance between them. The coefficient of proportion is determined when once the units are chosen, and as soon as we know the numerical values of this force, of the two masses, and of their distance. But when we wish to make laboratory experiments serious difficulties appear, owing to the weakness of the attraction between masses of ordinary dimensions. Microscopic forces, so to speak, have to be observed, and therefore all the causes of errors have to be avoided which would be unimportant in most other physical researches. It is known that Cavendish was the first who succeeded by means of the torsion balance in effecting fairly precise measurements. This method has been again taken in hand by different experimenters, and the most recent results are due to Mr Vernon Boys. This learned physicist is also the author of a most useful practical invention, and has succeeded in making quartz threads as fine as can be desired and extremely uniform. He finds that these threads possess valuable properties, such as perfect elasticity and great tenacity. He has been able, with threads not more than 1/500 of a millimetre in diameter, to measure with precision couples of an order formerly considered outside the range of experiment, and to reduce the dimensions of the apparatus of Cavendish in the proportion of 150 to 1. The great advantage found in the use of these small instruments is the better avoidance of the perturbations arising from draughts of air, and of the very serious influence of the slightest inequality in temperature.

Other methods have been employed in late years by other experimenters, such as the method of Baron Eötvös, founded on the use of a torsion lever, the method of the ordinary balance, used especially by Professors Richarz and Krigar-Menzel and also by Professor Poynting, and the method of M. Wilsing, who uses a balance with a vertical beam. The results fairly agree, and lead to attributing to the earth a density equal to 5.527.

The most familiar manifestation of gravitation is gravity. The action of the earth on the unit of mass placed in one point, and the intensity of gravity, is measured, as we know, by the aid of a pendulum. The methods of measurement, whether by absolute or by relative determinations, so greatly improved by Borda and Bessel, have been still further improved by various geodesians, among whom should be mentioned M. von Sterneek and General Defforges. Numerous observations have been made in all parts of the world by various explorers, and have led to a fairly complete knowledge of the distribution of gravity over the surface of the globe. Thus we have succeeded in making evident anomalies which would not easily find their place in the formula of Clairaut.

Another constant, the determination of which is of the greatest utility in astronomy of position, and the value of which enters into electromagnetic theory, has to-day assumed, with the new ideas on the constitution of matter, a still more considerable importance. I refer to the speed of light, which appears to us, as we shall see further on, the maximum value of speed which can be given to a material body.

After the historical experiments of Fizeau and Foucault, taken up afresh, as we know, partly by Cornu, and partly by Michelson and Newcomb, it remained still possible to increase the precision of the measurements. Professor Michelson has undertaken some new researches by a method which is a combination of the principle of the toothed wheel of Fizeau with the revolving mirror of Foucault. The toothed wheel is here replaced, however, by a grating, in which the lines and the spaces between them take the place of the teeth and the gaps, the reflected light only being returned when it strikes on the space between two lines. The illustrious American physicist estimates that he can thus evaluate to nearly five kilometres the path traversed by light in one second. This approximation corresponds to a relative value of a few hundred-thousandths, and it far exceeds those hitherto attained by the best experimenters. When all the experiments are completed, they will perhaps solve certain questions still in suspense; for instance, the question whether the speed of propagation depends on intensity. If this turns out to be the case, we should be brought to the important conclusion that the amplitude of the oscillations, which is certainly very small in relation to the already tiny wave-lengths, cannot be considered as unimportant in regard to these lengths. Such would seem to have been the result of the curious experiments of M. Muller and of M. Ebert, but these results have been recently disputed by M. Doubt.

In the case of sound vibrations, on the other hand, it should be noted that experiment, consistently with the theory, proves that the speed increases with the amplitude, or, if you will, with the intensity. M. Violle has published an important series of experiments on the speed of propagation of very condensed waves, on the deformations of these waves, and on the relations of the speed and the pressure, which verify in a remarkable manner the results foreshadowed by the already old calculations of Riemann, repeated later by Hugoniot. If, on the contrary, the amplitude is sufficiently small, there exists a speed limit which is the same in a large pipe and in free air. By some beautiful experiments, MM. Violle and Vautier have clearly shown that any disturbance in the air melts somewhat quickly into a single wave of given form, which is propagated to a distance, while gradually becoming weaker and showing a constant speed which differs little in dry air at 0° C. from 331.36 metres per second. In a narrow pipe the influence of the walls makes itself felt and produces various effects, in particular a kind of dispersion in space of the harmonics of the sound. This phenomenon, according to M. Brillouin, is perfectly explicable by a theory similar to the theory of gratings.

CHAPTER III
PRINCIPLES
§ 1. THE PRINCIPLES OF PHYSICS

Facts conscientiously observed lead by induction to the enunciation of a certain number of laws or general hypotheses which are the principles already referred to. These principal hypotheses are, in the eyes of a physicist, legitimate generalizations, the consequences of which we shall be able at once to check by the experiments from which they issue.

Among the principles almost universally adopted until lately figure prominently those of mechanics—such as the principle of relativity, and the principle of the equality of action and reaction. We will not detail nor discuss them here, but later on we shall have an opportunity of pointing out how recent theories on the phenomena of electricity have shaken the confidence of physicists in them and have led certain scholars to doubt their absolute value.

The principle of Lavoisier, or principle of the conservation of mass, presents itself under two different aspects according to whether mass is looked upon as the coefficient of the inertia of matter or as the factor which intervenes in the phenomena of universal attraction, and particularly in gravitation. We shall see when we treat of these theories, how we have been led to suppose that inertia depended on velocity and even on direction. If this conception were exact, the principle of the invariability of mass would naturally be destroyed. Considered as a factor of attraction, is mass really indestructible?

A few years ago such a question would have seemed singularly audacious. And yet the law of Lavoisier is so far from self-evident that for centuries it escaped the notice of physicists and chemists. But its great apparent simplicity and its high character of generality, when enunciated at the end of the eighteenth century, rapidly gave it such an authority that no one was able to any longer dispute it unless he desired the reputation of an oddity inclined to paradoxical ideas.

It is important, however, to remark that, under fallacious metaphysical appearances, we are in reality using empty words when we repeat the aphorism, "Nothing can be lost, nothing can be created," and deduce from it the indestructibility of matter. This indestructibility, in truth, is an experimental fact, and the principle depends on experiment. It may even seem, at first sight, more singular than not that the weight of a bodily system in a given place, or the quotient of this weight by that of the standard mass—that is to say, the mass of these bodies—remains invariable, both when the temperature changes and when chemical reagents cause the original materials to disappear and to be replaced by new ones. We may certainly consider that in a chemical phenomenon annihilations and creations of matter are really produced; but the experimental law teaches us that there is compensation in certain respects.

The discovery of the radioactive bodies has, in some sort, rendered popular the speculations of physicists on the phenomena of the disaggregation of matter. We shall have to seek the exact meaning which ought to be given to the experiments on the emanation of these bodies, and to discover whether these experiments really imperil the law of Lavoisier.

For some years different experimenters have also effected many very precise measurements of the weight of divers bodies both before and after chemical reactions between these bodies. Two highly experienced and cautious physicists, Professors Landolt and Heydweiller, have not hesitated to announce the sensational result that in certain circumstances the weight is no longer the same after as before the reaction. In particular, the weight of a solution of salts of copper in water is not the exact sum of the joint weights of the salt and the water. Such experiments are evidently very delicate; they have been disputed, and they cannot be considered as sufficient for conviction. It follows nevertheless that it is no longer forbidden to regard the law of Lavoisier as only an approximate law; according to Sandford and Ray, this approximation would be about 1/2,400,000. This is also the result reached by Professor Poynting in experiments regarding the possible action of temperature on the weight of a body; and if this be really so, we may reassure ourselves, and from the point of view of practical application may continue to look upon matter as indestructible.

The principles of physics, by imposing certain conditions on phenomena, limit after a fashion the field of the possible. Among these principles is one which, notwithstanding its importance when compared with that of universally known principles, is less familiar to some people. This is the principle of symmetry, more or less conscious applications of which can, no doubt, be found in various works and even in the conceptions of Copernican astronomers, but which was generalized and clearly enunciated for the first time by the late M. Curie. This illustrious physicist pointed out the advantage of introducing into the study of physical phenomena the considerations on symmetry familiar to crystallographers; for a phenomenon to take place, it is necessary that a certain dissymmetry should previously exist in the medium in which this phenomenon occurs. A body, for instance, may be animated with a certain linear velocity or a speed of rotation; it may be compressed, or twisted; it may be placed in an electric or in a magnetic field; it may be affected by an electric current or by one of heat; it may be traversed by a ray of light either ordinary or polarized rectilineally or circularly, etc.:—in each case a certain minimum and characteristic dissymmetry is necessary at every point of the body in question.

This consideration enables us to foresee that certain phenomena which might be imagined *a priori* cannot exist. Thus, for instance, it is impossible that an electric field, a magnitude directed and not superposable on its image in a mirror perpendicular to its direction, could be created at right angles to the plane of symmetry of the medium; while it would be possible to create a magnetic field under the same conditions.

This consideration thus leads us to the discovery of new phenomena; but it must be understood that it cannot of itself give us absolutely precise notions as to the nature of these phenomena, nor disclose their order of magnitude.

§ 2. THE PRINCIPLE OF THE CONSERVATION OF ENERGY

Dominating not physics alone, but nearly every other science, the principle of the conservation of energy is justly considered as the grandest conquest of contemporary thought. It shows us in a powerful light the most diverse questions; it introduces order into the most varied studies; it leads to a clear and coherent interpretation of phenomena which, without it, appear to have no connexion with each other; and it supplies precise and exact numerical relations between the magnitudes which enter into these phenomena.

The boldest minds have an instinctive confidence in it, and it is the principle which has most stoutly resisted that assault which the daring of a few theorists has lately directed to the overthrow of the general principles of physics. At every new discovery, the first thought of physicists is to find out how it accords with the principle of the conservation of energy. The application of the principle, moreover, never fails to give valuable hints on the new phenomenon, and often even suggests a complementary discovery. Up till now it seems never to have received a check, even the extraordinary properties of radium not seriously contradicting it; also the general form in which it is enunciated gives it such a suppleness that it is no doubt very difficult to overthrow.

I do not claim to set forth here the complete history of this principle, but I will endeavour to show with what pains it was born, how it was kept back in its early days and then obstructed in its development by the unfavourable conditions of the surroundings in which it appeared. It first of all came, in fact, to oppose itself to the reigning theories; but, little by little, it acted on these theories, and they were modified under its pressure; then, in their turn, these theories reacted on it and changed its primitive form.

It had to be made less wide in order to fit into the classic frame, and was absorbed by mechanics; and if it thus became less general, it gained in precision what it lost in extent. When once definitely admitted and classed, as it were, in the official domain of science, it endeavoured to burst its bonds and return to a more independent and larger life. The history of this principle is similar to that of all evolutions.

It is well known that the conservation of energy was, at first, regarded from the point of view of the reciprocal transformations between heat and work, and that the principle received its first clear enunciation in the particular case of the principle of equivalence. It is, therefore, rightly considered that the scholars who were the first to doubt the material nature of caloric were the precursors of R. Mayer; their ideas, however, were the same as those of the celebrated German doctor, for they sought especially to demonstrate that heat was a mode of motion.

Without going back to early and isolated attempts like those of Daniel Bernoulli, who, in his hydrodynamics, propounded the basis of the kinetic theory of gases, or the researches of Boyle on friction, we may recall, to show how it was propounded in former times, a rather forgotten page of the *Mémoire sur la Chaleur*, published in 1780 by Lavoisier and Laplace: "Other physicists," they wrote, after setting out the theory of caloric, "think that heat is nothing but the result of the insensible vibrations of matter.... In the system we are now examining, heat is the *vis viva* resulting from the insensible movements of the molecules of a body; it is the sum of the products of the mass of each molecule by the square of its velocity.... We shall not decide between the two preceding hypotheses; several phenomena seem to support the last mentioned—for instance, that of the heat produced by the friction of two solid bodies. But there are others which are more simply explained by the first, and perhaps they both operate at once." Most of the physicists of that period, however, did not share the prudent doubts of Lavoisier and Laplace. They admitted, without hesitation, the first hypothesis; and, four years after the appearance of the *Mémoire sur la Chaleur*, Sigaud de Lafond, a professor of physics of great reputation, wrote: "Pure Fire, free from all state of combination, seems to be an assembly of particles of a simple, homogeneous, and absolutely unalterable matter, and all the properties of this element indicate that these particles are infinitely small and free, that they have no sensible cohesion, and that they are moved in every possible direction by a continual and rapid motion which is essential to them.... The extreme tenacity and the surprising mobility of its molecules are manifestly shown by the ease with which it penetrates into the most compact bodies and by its tendency to put itself in equilibrium throughout all bodies near to it."

It must be acknowledged, however, that the idea of Lavoisier and Laplace was rather vague and even inexact on one important point. They admitted it to be evident that "all variations of heat, whether real or apparent, undergone by a bodily system when changing its state, are produced in inverse order when the system passes back to its original state." This phrase is the very denial of equivalence where these changes of state are accompanied by external work.

Laplace, moreover, himself became later a very convinced partisan of the hypothesis of the material nature of caloric, and his immense authority, so fortunate in other respects for the development of science, was certainly in this case the cause of the retardation of progress.

The names of Young, Rumford, Davy, are often quoted among those physicists who, at the commencement of the nineteenth century, caught sight of the new truths as to the nature of heat. To these names is very properly added that of Sadi Carnot. A note found among his papers unquestionably proves that, before 1830, ideas had occurred to him from which it resulted that in producing work an equivalent amount of heat was destroyed. But the year 1842 is particularly memorable in the history of science as the year in which Jules Robert Mayer succeeded, by an entirely personal effort, in really enunciating the principle of the conservation of energy. Chemists recall with just pride that the *Remarques sur les forces de la nature animée*, contemptuously rejected by all the journals of physics, were received and published in the *Annalen* of Liebig. We ought never to forget this example, which shows with what difficulty a new idea contrary to the classic theories of the period succeeds in coming to the front; but extenuating circumstances may be urged on behalf of the physicists.

Robert Mayer had a rather insufficient mathematical education, and his Memoirs, the *Remarques*, as well as the ulterior publications, *Mémoire sur le mouvement organique et la nutrition* and the *Matériaux pour la dynamique du ciel*, contain, side by side with very profound ideas, evident errors in mechanics. Thus it often happens that discoveries put forward in a somewhat

56

vague manner by adventurous minds not overburdened by the heavy baggage of scientific erudition, who audaciously press forward in advance of their time, fall into quite intelligible oblivion until rediscovered, clarified, and put into shape by slower but surer seekers. This was the case with the ideas of Mayer. They were not understood at first sight, not only on account of their originality, but also because they were couched in incorrect language.

Mayer was, however, endowed with a singular strength of thought; he expressed in a rather confused manner a principle which, for him, had a generality greater than mechanics itself, and so his discovery was in advance not only of his own time but of half the century. He may justly be considered the founder of modern energetics.

Freed from the obscurities which prevented its being clearly perceived, his idea stands out to-day in all its imposing simplicity. Yet it must be acknowledged that if it was somewhat denaturalised by those who endeavoured to adapt it to the theories of mechanics, and if it at first lost its sublime stamp of generality, it thus became firmly fixed and consolidated on a more stable basis.

The efforts of Helmholtz, Clausius, and Lord Kelvin to introduce the principle of the conservation of energy into mechanics, were far from useless. These illustrious physicists succeeded in giving a more precise form to its numerous applications; and their attempts thus contributed, by reaction, to give a fresh impulse to mechanics, and allowed it to be linked to a more general order of facts. If energetics has not been able to be included in mechanics, it seems indeed that the attempt to include mechanics in energetics was not in vain.

In the middle of the last century, the explanation of all natural phenomena seemed more and more referable to the case of central forces. Everywhere it was thought that reciprocal actions between material points could be perceived, these points being attracted or repelled by each other with an intensity depending only on their distance or their mass. If, to a system thus composed, the laws of the classical mechanics are applied, it is shown that half the sum of the product of the masses by the square of the velocities, to which is added the work which might be accomplished by the forces to which the system would be subject if it returned from its actual to its initial position, is a sum constant in quantity.

This sum, which is the mechanical energy of the system, is therefore an invariable quantity in all the states to which it may be brought by the interaction of its various parts, and the word energy well expresses a capital property of this quantity. For if two systems are connected in such a way that any change produced in the one necessarily brings about a change in the other, there can be no variation in the characteristic quantity of the second except so far as the characteristic quantity of the first itself varies—on condition, of course, that the connexions are made in such a manner as to introduce no new force. It will thus be seen that this quantity well expresses the capacity possessed by a system for modifying the state of a neighbouring system to which we may suppose it connected.

Now this theorem of pure mechanics was found wanting every time friction took place—that is to say, in all really observable cases. The more perceptible the friction, the more considerable the difference; but, in addition, a new phenomenon always appeared and heat was produced. By experiments which are now classic, it became established that the quantity of heat thus created independently of the nature of the bodies is always (provided no other phenomena intervene) proportional to the energy which has disappeared. Reciprocally, also, heat may disappear, and we always find a constant relation between the quantities of heat and work which mutually replace each other.

It is quite clear that such experiments do not prove that heat is work. We might just as well say that work is heat. It is making a gratuitous hypothesis to admit this reduction of heat to mechanism; but this hypothesis was so seductive, and so much in conformity with the desire of nearly all physicists to arrive at some sort of unity in nature, that they made it with eagerness and became unreservedly convinced that heat was an active internal force.

Their error was not in admitting this hypothesis; it was a legitimate one since it has proved very fruitful. But some of them committed the fault of forgetting that it was an hypothesis, and considered it a demonstrated truth. Moreover, they were thus brought to see in phenomena nothing but these two particular forms of energy which in their minds were easily identified with each other.

From the outset, however, it became manifest that the principle is applicable to cases where heat plays only a parasitical part. There were thus discovered, by translating the principle of equivalence, numerical relations between the magnitudes of electricity, for instance, and the magnitudes of mechanics. Heat was a sort of variable intermediary convenient for calculation, but introduced in a roundabout way and destined to disappear in the final result.

Verdet, who, in lectures which have rightly remained celebrated, defined with remarkable clearness the new theories, said, in 1862: "Electrical phenomena are always accompanied by calorific manifestations, of which the study belongs to the mechanical theory of heat. This study, moreover, will not only have the effect of making known to us interesting facts in electricity, but will throw some light on the phenomena of electricity themselves."

The eminent professor was thus expressing the general opinion of his contemporaries, but he certainly seemed to have felt in advance that the new theory was about to penetrate more deeply into the inmost nature of things. Three years previously, Rankine also had put forth some very remarkable ideas the full meaning of which was not at first well understood. He it was who comprehended the utility of employing a more inclusive term, and invented the phrase energetics. He also endeavoured to create a new doctrine of which rational mechanics should be only a particular case; and he showed that it was possible to abandon the ideas of atoms and central forces, and to construct a more general system by substituting for the ordinary consideration of forces that of the energy which exists in all bodies, partly in an actual, partly in a potential state.

By giving more precision to the conceptions of Rankine, the physicists of the end of the nineteenth century were brought to consider that in all physical phenomena there occur apparitions and disappearances which are balanced by various energies. It is natural, however, to suppose that these equivalent apparitions and disappearances correspond to transformations and not to simultaneous creations and destructions. We thus represent energy to ourselves as taking different forms—mechanical, electrical, calorific, and chemical—capable of changing one into the other, but in such a way that the quantitative value always remains the same. In like manner a bank draft may be represented by notes, gold, silver, or bullion. The earliest known form of energy, *i.e.* work, will serve as the standard as gold serves as the monetary standard, and energy in all its forms will be estimated by the corresponding work. In each particular case we can strictly define and measure, by the correct application of the principle of the conservation of energy, the quantity of energy evolved under a given form.

We can thus arrange a machine comprising a body capable of evolving this energy; then we can force all the organs of this machine to complete an entirely closed cycle, with the exception of the body itself, which, however, has to return to such a state that all the variables from which this state depends resume their initial values except the particular variable to which the evolution of the energy under consideration is linked. The difference between the work thus accomplished and that which would have been obtained if this variable also had returned to its original value, is the measure of the energy evolved.

In the same way that, in the minds of mechanicians, all forces of whatever origin, which are capable of compounding with each other and of balancing each other, belong to the same category of beings, so for many physicists energy is a sort of entity which we find under various aspects. There thus exists for them a world, which comes in some way to superpose itself upon the world of matter—that is to say, the world of energy, dominated in its turn by a fundamental law similar to that of Lavoisier. [5] This conception, as we have already seen, passes the limit of experience; but others go further still. Absorbed in the contemplation of this new world, they succeed in persuading themselves that the old world of matter has no real existence and that energy is sufficient by itself to give us a complete comprehension of the Universe and of all the phenomena produced in it. They point out that all our sensations correspond to changes of energy, and that everything apparent to our senses is, in truth, energy. The famous experiment of the blows with a stick by which it was demonstrated to a sceptical philosopher that an outer world existed, only proves, in reality, the existence of energy, and not that of matter. The stick in itself is inoffensive, as Professor Ostwald remarks, and it is its *vis viva*, its kinetic energy, which is painful to us; while if we possessed a speed equal to its own, moving in the same direction, it would no longer exist so far as our sense of touch is concerned.

On this hypothesis, matter would only be the capacity for kinetic energy, its pretended impenetrability energy of volume, and its weight energy of position in the particular form which presents itself in universal gravitation; nay, space itself would only be known to us by the expenditure of energy necessary to penetrate it. Thus in all physical phenomena we should only have to regard the quantities of energy brought into play, and all the equations which link the phenomena to one another would have no meaning but when they apply to exchanges of energy. For energy alone can be common to all phenomena.

This extreme manner of regarding things is seductive by its originality, but appears somewhat insufficient if, after enunciating generalities, we look more closely into the question. From the philosophical point of view it may, moreover, seem difficult not to conclude, from the qualities which reveal, if you will, the varied forms of energy, that there exists a substance possessing these qualities. This energy, which resides in one region, and which transports itself from one spot to another, forcibly brings to mind, whatever view we may take of it, the idea of matter.

Helmholtz endeavoured to construct a mechanics based on the idea of energy and its conservation, but he had to invoke a second law, the principle of least action. If he thus succeeded in dispensing with the hypothesis of atoms, and in showing that the new mechanics gave us to understand the impossibility of certain movements which, according to the old, ought to have been but never were experimentally produced, he was only able to do so because the principle of least action necessary for his theory became evident in the case of those irreversible phenomena which alone really exist in Nature. The energetists have thus not succeeded in forming a thoroughly sound system, but their efforts have at all events been partly successful. Most physicists are of their opinion, that kinetic energy is only a particular variety of energy to which we have no right to wish to connect all its other forms.

If these forms showed themselves to be innumerable throughout the Universe, the principle of the conservation of energy would, in fact, lose a great part of its importance. Every time that a certain quantity of energy seemed to appear or disappear, it would always be permissible to suppose that an equivalent quantity had appeared or disappeared somewhere else under a new form; and thus the principle would in a way vanish. But the known forms of energy are fairly restricted in number, and the necessity of recognising new ones seldom makes itself felt. We shall see, however, that to explain, for instance, the paradoxical properties of radium and to re-establish concord between these properties and the principle of the conservation of energy, certain physicists have recourse to the hypothesis that radium borrows an unknown energy from the medium in which it is plunged. This hypothesis, however, is in no way necessary; and in a few other rare cases in which similar hypotheses have had to be set up, experiment has always in the long run enabled us to discover some phenomenon which had escaped the first observers and which corresponds exactly to the variation of energy first made evident.

One difficulty, however, arises from the fact that the principle ought only to be applied to an isolated system. Whether we imagine actions at a distance or believe in intermediate media, we must always recognise that there exist no bodies in the world incapable of acting on each other, and we can never affirm that some modification in the energy of a given place may not have its echo in some unknown spot afar off. This difficulty may sometimes render the value of the principle rather illusory.

Similarly, it behoves us not to receive without a certain distrust the extension by certain philosophers to the whole Universe, of a property demonstrated for those restricted systems which observation can alone reach. We know nothing of the Universe as a whole, and every generalization of this kind outruns in a singular fashion the limit of experiment.

Even reduced to the most modest proportions, the principle of the conservation of energy retains, nevertheless, a paramount importance; and it still preserves, if you will, a high philosophical value. M.J. Perrin justly points out that it gives us a form under which we are experimentally able to grasp causality, and that it teaches us that a result has to be purchased at the cost of a determined effort.

We can, in fact, with M. Perrin and M. Langevin, represent this in a way which puts this characteristic in evidence by enunciating it as follows: "If at the cost of a change C we can obtain a change K, there will never be acquired at the same cost, whatever the mechanism employed, first the change K and in addition some other change, unless this latter be one that is otherwise known to cost nothing to produce or to destroy." If, for instance, the fall of a weight can be accompanied, without anything else being produced, by another transformation—the melting of a certain mass of ice, for example—it will be impossible, no matter how you set about it or whatever the mechanism used, to associate this same transformation with the melting of another weight of ice.

We can thus, in the transformation in question, obtain an appropriate number which will sum up that which may be expected from the external effect, and can give, so to speak, the price at which this transformation is bought, measure its invariable value by a common measure (for instance, the melting of the ice), and, without any ambiguity, define the energy lost during the transformation as proportional to the mass of ice which can be associated with it. This measure is, moreover, independent of the particular phenomenon taken as the common measure.

§ 3. THE PRINCIPLE OF CARNOT AND CLAUSIUS

The principle of Carnot, of a nature analogous to the principle of the conservation of energy, has also a similar origin. It was first enunciated, like the last named, although prior to it in time, in consequence of considerations which deal only with heat and mechanical work. Like it, too, it has evolved, grown, and invaded the entire domain of physics. It may be interesting to examine rapidly the various phases of this evolution. The origin of the principle of Carnot is clearly determined, and it is very rare to be able to go back thus certainly to the source of a discovery. Sadi Carnot had, truth to say, no precursor. In his time heat engines were not yet very common, and no one had reflected much on their theory. He was doubtless the first to propound to himself certain questions, and certainly the first to solve them.

It is known how, in 1824, in his *Réflexions sur la puissance motrice du feu*, he endeavoured to prove that "the motive power of heat is independent of the agents brought into play for its realization," and that "its quantity is fixed solely by the temperature of the bodies between which, in the last resort, the transport of caloric is effected"—at least in all engines in which "the method of developing the motive power attains the perfection of which it is capable"; and this is, almost textually, one of the enunciations of the principle at the present day. Carnot perceived very clearly the great fact that, to produce work by heat, it is necessary to have at one's disposal a fall of temperature. On this point he expresses himself with perfect clearness: "The motive power of a fall of water depends on its height and on the quantity of liquid; the motive power of heat depends also on the quantity of caloric employed, and on what might be called—in fact, what we shall call—the height of fall, that is to say, the difference in temperature of the bodies between which the exchange of caloric takes place."

Starting with this idea, he endeavours to demonstrate, by associating two engines capable of working in a reversible cycle, that the principle is founded on the impossibility of perpetual motion.

His memoir, now celebrated, did not produce any great sensation, and it had almost fallen into deep oblivion, which, in consequence of the discovery of the principle of equivalence, might have seemed perfectly justified. Written, in fact, on the hypothesis of the indestructibility of caloric, it was to be expected that this memoir should be condemned in the name of the new doctrine, that is, of the principle recently brought to light.

It was really making a new discovery to establish that Carnot's fundamental idea survived the destruction of the hypothesis on the nature of heat, on which he seemed to rely. As he no doubt himself perceived, his idea was quite independent of this hypothesis, since, as we have seen, he was led to surmise that heat could disappear; but his demonstrations needed to be recast and, in some points, modified.

It is to Clausius that was reserved the credit of rediscovering the principle, and of enunciating it in language conformable to the new doctrines, while giving it a much greater generality. The postulate arrived at by experimental induction, and which must be admitted without demonstration, is, according to Clausius, that in a series of transformations in which the final is identical with the initial stage, it is impossible for heat to pass from a colder to a warmer body unless some other accessory phenomenon occurs at the same time.

Still more correctly, perhaps, an enunciation can be given of the postulate which, in the main, is analogous, by saying: A heat motor, which after a series of transformations returns to its initial state, can only furnish work if there exist at least two sources of heat, and if a certain quantity of heat is given to one of the sources, which can never be the hotter of the two. By the expression "source of heat," we mean a body exterior to the system and capable of furnishing or withdrawing heat from it.

Starting with this principle, we arrive, as does Clausius, at the demonstration that the output of a reversible machine working between two given temperatures is greater than that of any non-reversible engine, and that it is the same for all reversible machines working between these two temperatures.

This is the very proposition of Carnot; but the proposition thus stated, while very useful for the theory of engines, does not yet present any very general interest. Clausius, however, drew from it much more important consequences. First, he showed that the principle conduces to the definition of an absolute scale of temperature; and then he was brought face to face with a new notion which allows a strong light to be thrown on the questions of physical equilibrium. I refer to entropy.

It is still rather difficult to strip entirely this very important notion of all analytical adornment. Many physicists hesitate to utilize it, and even look upon it with some distrust, because they see in it a purely mathematical function without any definite physical meaning. Perhaps they are here unduly severe, since they often admit too easily the objective existence of quantities which they cannot define. Thus, for instance, it is usual almost every day to speak of the heat possessed by a body. Yet no body in reality possesses a definite quantity of heat even relatively to any initial state; since starting from this point of departure, the quantities of heat it may have gained or lost vary with the road taken and even with the means employed to follow it. These expressions of heat gained or lost are, moreover, themselves evidently incorrect, for heat can no longer be considered as a sort of fluid passing from one body to another.

The real reason which makes entropy somewhat mysterious is that this magnitude does not fall directly under the ken of any of our senses; but it possesses the true characteristic of a concrete physical magnitude, since it is, in principle at least, measurable. Various authors of thermodynamical researches, amongst whom M. Mouret should be particularly mentioned, have endeavoured to place this characteristic in evidence.

Consider an isothermal transformation. Instead of leaving the heat abandoned by the body subjected to the transformation—water condensing in a state of saturated vapour, for instance—to pass directly into an ice calorimeter, we can transmit this heat to the calorimeter by the intermediary of a reversible Carnot engine. The engine having absorbed this quantity of heat, will only give back to the ice a lesser quantity of heat; and the weight of the melted ice, inferior to that which might have been directly given back, will serve as a measure of the isothermal transformation thus effected. It can be easily shown that this measure is independent of the apparatus used. It consequently becomes a numerical element characteristic of the body considered, and is called its entropy. Entropy, thus defined, is a variable which, like pressure or volume, might serve concurrently with another variable, such as pressure or volume, to define the state of a body.

It must be perfectly understood that this variable can change in an independent manner, and that it is, for instance, distinct from the change of temperature. It is also distinct from the change which consists in losses or gains of heat. In chemical reactions, for example, the entropy increases without the substances borrowing any heat. When a perfect gas dilates in a vacuum its entropy increases, and yet the temperature does not change, and the gas has neither been able to give nor receive heat. We thus come to conceive that a physical phenomenon cannot be considered known to us if the variation of entropy is not given, as are the variations of temperature and of pressure or the exchanges of heat. The change of entropy is, properly speaking, the most characteristic fact of a thermal change.

It is important, however, to remark that if we can thus easily define and measure the difference of entropy between two states of the same body, the value found depends on the state arbitrarily chosen as the zero point of entropy; but this is not a very serious difficulty, and is analogous to that which occurs in the evaluation of other physical magnitudes—temperature, potential, etc.

A graver difficulty proceeds from its not being possible to define a difference, or an equality, of entropy between two bodies chemically different. We are unable, in fact, to pass by any means, reversible or not, from one to the other, so long as the transmutation of matter is regarded as impossible; but it is well understood that it is nevertheless possible to compare the variations of entropy to which these two bodies are both of them individually subject.

Neither must we conceal from ourselves that the definition supposes, for a given body, the possibility of passing from one state to another by a reversible transformation. Reversibility is an ideal and extreme case which cannot be realized, but which can be approximately attained in many circumstances. So with gases and with perfectly elastic bodies, we effect sensibly reversible transformations, and changes of physical state are practically reversible. The discoveries of Sainte-Claire Deville have brought many chemical phenomena into a similar category, and reactions such as solution, which used to be formerly the type of an irreversible phenomenon, may now often be effected by sensibly reversible means. Be that as it may, when once the definition is admitted, we arrive, by taking as a basis the principles set forth at the inception, at the demonstration of the celebrated theorem of Clausius: *The entropy of a thermally isolated system continues to increase incessantly.*

It is very evident that the theorem can only be worth applying in cases where the entropy can be exactly defined; but, even when thus limited, the field still remains vast, and the harvest which we can there reap is very abundant.

Entropy appears, then, as a magnitude measuring in a certain way the evolution of a system, or, at least, as giving the direction of this evolution. This very important consequence certainly did not escape Clausius, since the very name of entropy, which he chose to designate this magnitude, itself signifies evolution. We have succeeded in defining this entropy by demonstrating, as has been said, a certain number of propositions which spring from the postulate of Clausius; it is, therefore, natural to suppose that this postulate itself contains *in potentia* the very idea of a necessary evolution of physical systems. But as it was first enunciated, it contains it in a deeply hidden way.

No doubt we should make the principle of Carnot appear in an interesting light by endeavouring to disengage this fundamental idea, and by placing it, as it were, in large letters. Just as, in elementary geometry, we can replace the postulate of Euclid by other equivalent propositions, so the postulate of thermodynamics is not necessarily fixed, and it is instructive to try to give it the most general and suggestive character.

MM. Perrin and Langevin have made a successful attempt in this direction. M. Perrin enunciates the following principle: *An isolated system never passes twice through the same state*. In this form, the principle affirms that there exists a necessary order in the succession of two phenomena; that evolution takes place in a determined direction. If you prefer it, it may be thus stated: *Of two converse transformations unaccompanied by any external effect, one only is possible*. For instance, two gases may diffuse themselves one in the other in constant volume, but they could not conversely separate themselves spontaneously.

Starting from the principle thus put forward, we make the logical deduction that one cannot hope to construct an engine which should work for an indefinite time by heating a hot source and by cooling a cold one. We thus come again into the route traced by Clausius, and from this point we may follow it strictly.

Whatever the point of view adopted, whether we regard the proposition of M. Perrin as the corollary of another experimental postulate, or whether we consider it as a truth which we admit *a priori* and verify through its consequences, we are led to consider that in its entirety the principle of Carnot resolves itself into the idea that we cannot go back along the course of life, and that the evolution of a system must follow its necessary progress.

Clausius and Lord Kelvin have drawn from these considerations certain well-known consequences on the evolution of the Universe. Noticing that entropy is a property added to matter, they admit that there is in the world a total amount of entropy; and as all real changes which are produced in any system correspond to an increase of entropy, it may be said that the entropy of the world is continually increasing. Thus the quantity of energy existing in the Universe remains constant, but transforms itself little by little into heat uniformly distributed at a temperature everywhere identical. In the end, therefore, there will be neither chemical phenomena nor manifestation of life; the world will still exist, but without motion, and, so to speak, dead.

These consequences must be admitted to be very doubtful; we cannot in any certain way apply to the Universe, which is not a finite system, a proposition demonstrated, and that not unreservedly, in the sharply limited case of a finite system. Herbert Spencer, moreover, in his book on *First Principles*, brings out with much force the idea that, even if the Universe came to an end, nothing would allow us to conclude that, once at rest, it would remain so indefinitely. We may recognise that the state in which we are began at the end of a former evolutionary period, and that the end of the existing era will mark the beginning of a new one.

Like an elastic and mobile object which, thrown into the air, attains by degrees the summit of its course, then possesses a zero velocity and is for a moment in equilibrium, and then falls on touching the ground to rebound, so the world should be subjected to huge oscillations which first bring it to a maximum of entropy till the moment when there should be produced a slow evolution in the contrary direction bringing it back to the state from which it started. Thus, in the infinity of time, the life of the Universe proceeds without real stop.

This conception is, moreover, in accordance with the view certain physicists take of the principle of Carnot. We shall see, for example, that in the kinetic theory we are led to admit that, after waiting sufficiently long, we can witness the return of the various states through which a mass of gas, for example, has passed in its series of transformations.

If we keep to the present era, evolution has a fixed direction—that which leads to an increase of entropy; and it is possible to enquire, in any given system to what physical manifestations this increase corresponds. We note that kinetic, potential, electrical, and chemical forms of energy have a great tendency to transform themselves into calorific energy. A chemical reaction, for example, gives out energy; but if the reaction is not produced under very special conditions, this energy immediately passes into the calorific form. This is so true, that chemists currently speak of the heat given out by reactions instead of regarding the energy disengaged in general.

In all these transformations the calorific energy obtained has not, from a practical point of view, the same value at which it started. One cannot, in fact, according to the principle of Carnot, transform it integrally into mechanical energy, since the heat possessed by a body can only yield work on condition that a part of it falls on a body with a lower temperature. Thus appears the idea that energies which exchange with each other and correspond to equal quantities have not the same qualitative value. Form has its importance, and there are persons who prefer a golden louis to four pieces of five francs. The principle of Carnot would thus lead us to consider a certain classification of energies, and would show us that, in the transformations possible, these energies always tend to a sort of diminution of quality—that is, to a *degradation*.

It would thus reintroduce an element of differentiation of which it seems very difficult to give a mechanical explanation. Certain philosophers and physicists see in this fact a reason which condemns *a priori* all attempts made to give a mechanical explanation of the principle of Carnot.

It is right, however, not to exaggerate the importance that should be attributed to the phrase degraded energy. If the heat is not equivalent to the work, if heat at 99° is not equivalent to heat at 100°, that means that we cannot in practice construct an engine which shall transform all this heat into work, or that, for the same cold source, the output is greater when the temperature of the hot source is higher; but if it were possible that this cold source had itself the temperature of absolute zero, the whole heat would reappear in the form of work. The case here considered is an ideal and extreme case, and we naturally cannot realize it; but this consideration suffices to make it plain that the classification of energies is a little arbitrary and depends more, perhaps, on the conditions in which mankind lives than on the inmost nature of things.

In fact, the attempts which have often been made to refer the principle of Carnot to mechanics have not given convincing results. It has nearly always been necessary to introduce into the attempt some new hypothesis independent of the fundamental hypotheses of ordinary mechanics, and equivalent, in reality, to one of the postulates on which the ordinary exposition of the second law of thermodynamics is founded. Helmholtz, in a justly celebrated theory, endeavoured to fit the principle of Carnot into the principle of least action; but the difficulties regarding the mechanical interpretation of the irreversibility of physical phenomena remain entire. Looking at the question, however, from the point of view at which the partisans of the kinetic theories of matter place themselves, the principle is viewed in a new aspect. Gibbs and afterwards Boltzmann and Professor Planck have put forward some very interesting ideas on this subject. By following the route they have traced, we come to consider the principle as pointing out to us that a given system tends towards the configuration presented by the maximum probability, and, numerically, the entropy would even be the logarithm of this probability. Thus two different gaseous masses, enclosed in two separate receptacles which have just been placed in communication, diffuse themselves one through the other, and it is highly improbable that, in their mutual shocks, both kinds of molecules should take a distribution of velocities which reduce them by a spontaneous phenomenon to the initial state.

We should have to wait a very long time for so extraordinary a concourse of circumstances, but, in strictness, it would not be impossible. The principle would only be a law of probability. Yet this probability is all the greater the more considerable is the number of molecules itself. In the phenomena habitually dealt with, this number is such that, practically, the variation of entropy in a constant sense takes, so to speak, the character of absolute certainty.

But there may be exceptional cases where the complexity of the system becomes insufficient for the application of the principle of Carnot;—as in the case of the curious movements of small particles suspended in a liquid which are known by the name of Brownian movements and can be observed under the microscope. The agitation here really seems, as M. Gouy has remarked, to be produced and continued indefinitely, regardless of any difference in temperature; and we seem to witness the incessant motion, in an isothermal medium, of the particles which constitute matter. Perhaps, however, we find ourselves already in conditions where the too great simplicity of the distribution of the molecules deprives the principle of its value.

M. Lippmann has in the same way shown that, on the kinetic hypothesis, it is possible to construct such mechanisms that we can so take cognizance of molecular movements that *vis viva* can be taken from them. The mechanisms of M. Lippmann are not, like the celebrated apparatus at one time devised by Maxwell, purely hypothetical. They do not suppose a partition with a hole impossible to be bored through matter where the molecular spaces would be larger than the hole itself. They have finite dimensions. Thus M. Lippmann considers a vase full of oxygen at a constant temperature. In the interior of this vase is placed a small copper ring, and the whole is set in a magnetic field. The oxygen molecules are, as we know, magnetic, and when passing through the interior of the ring they produce in this ring an induced current. During this time, it is true, other molecules emerge from the space enclosed by the circuit; but the two effects do not counterbalance each other, and the resulting current is maintained. There is elevation of temperature in the circuit in accordance with Joule's law; and this phenomenon, under such conditions, is incompatible with the principle of Carnot.

It is possible—and that, I think, is M. Lippmann's idea—to draw from his very ingenious criticism an objection to the kinetic theory, if we admit the absolute value of the principle; but we may also suppose that here again we are in presence of a system where the prescribed conditions diminish the complexity and render it, consequently, less probable that the evolution is always effected in the same direction.

In whatever way you look at it, the principle of Carnot furnishes, in the immense majority of cases, a very sure guide in which physicists continue to have the most entire confidence.

§ 4. THERMODYNAMICS

To apply the two fundamental principles of thermodynamics, various methods may be employed, equivalent in the main, but presenting as the cases vary a greater or less convenience.

In recording, with the aid of the two quantities, energy and entropy, the relations which translate analytically the two principles, we obtain two relations between the coefficients which occur in a given phenomenon; but it may be easier and also more suggestive to employ various functions of these quantities. In a memoir, of which some extracts appeared as early as 1869, a modest scholar, M. Massieu, indicated in particular a remarkable function which he termed a characteristic function, and by the employment of which calculations are simplified in certain cases.

In the same way J.W. Gibbs, in 1875 and 1878, then Helmholtz in 1882, and, in France, M. Duhem, from the year 1886 onward, have published works, at first ill understood, of which the renown was, however, considerable in the sequel, and in which they made use of analogous functions under the names of available energy, free energy, or internal thermodynamic potential. The magnitude thus designated, attaching, as a consequence of the two principles, to all states of the system, is perfectly determined when the temperature and other normal variables are known. It allows us, by calculations often very easy, to fix the conditions necessary and sufficient for the maintenance of the system in equilibrium by foreign bodies taken at the same temperature as itself.

One may hope to constitute in this way, as M. Duhem in a long and remarkable series of operations has specially endeavoured to do, a sort of general mechanics which will enable questions of statics to be treated with accuracy, and all the conditions of equilibrium of the system, including the calorific properties, to be determined. Thus, ordinary statics teaches us that a liquid with its vapour on the top forms a system in equilibrium, if we apply to the two fluids a pressure depending on temperature alone. Thermodynamics will furnish us, in addition, with the expression of the heat of vaporization and of, the specific heats of the two saturated fluids.

This new study has given us also most valuable information on compressible fluids and on the theory of elastic equilibrium. Added to certain hypotheses on electric or magnetic phenomena, it gives a coherent whole from which can be deduced the conditions of electric or magnetic equilibrium; and it illuminates with a brilliant light the calorific laws of electrolytic phenomena.

But the most indisputable triumph of this thermodynamic statics is the discovery of the laws which regulate the changes of physical state or of chemical constitution. J.W. Gibbs was the author of this immense progress. His memoir, now celebrated, on "the equilibrium of heterogeneous substances," concealed in 1876 in a review at that time of limited circulation, and rather heavy to read, seemed only to contain algebraic theorems applicable with difficulty to reality. It is known that Helmholtz independently succeeded, a few years later, in introducing thermodynamics into the domain of chemistry by his conception of the division of energy into free and into bound energy: the first, capable of undergoing all transformations, and particularly of transforming itself into external action; the second, on the other hand, bound, and only manifesting itself by giving out heat. When we measure chemical energy, we ordinarily let it fall wholly into the calorific form; but, in reality, it itself includes both parts, and it is the variation of the free energy and not that of the total energy measured by the integral disengagement of heat, the sign of which determines the direction in which the reactions are effected.

But if the principle thus enunciated by Helmholtz as a consequence of the laws of thermodynamics is at bottom identical with that discovered by Gibbs, it is more difficult of application and is presented under a more mysterious aspect. It was not until M. Van der Waals exhumed the memoir of Gibbs, when numerous physicists or chemists, most of them Dutch—Professor Van t'Hoff, Bakhius Roozeboom, and others—utilized the rules set forth in this memoir for the discussion of the most complicated chemical reactions, that the extent of the new laws was fully understood.

The chief rule of Gibbs is the one so celebrated at the present day under the name of the Phase Law. We know that by phases are designated the homogeneous substances into which a system is divided; thus carbonate of lime, lime, and carbonic acid gas are the three phases of a system which comprises Iceland spar partially dissociated into lime and carbonic acid gas. The number of phases added to the number of independent components—that is to say, bodies whose mass is left arbitrary by the chemical formulas of the substances entering into the reaction—fixes the general form of the law of equilibrium of the system; that is to say, the number of quantities which, by their variations (temperature and pressure), would be of a nature to modify its equilibrium by modifying the constitution of the phases.

Several authors, M. Raveau in particular, have indeed given very simple demonstrations of this law which are not based on thermodynamics; but thermodynamics, which led to its discovery, continues to give it its true scope. Moreover, it would not suffice merely to determine quantitatively those laws of which it makes known the general form. We must, if we wish to penetrate deeper into details, particularize the hypothesis, and admit, for instance, with Gibbs that we are dealing with perfect gases; while, thanks to thermodynamics, we can constitute a complete theory of dissociation which leads to formulas in complete accord with the numerical results of the experiment. We can thus follow closely all questions concerning the displacements of the equilibrium, and find a relation of the first importance between the masses of the bodies which react in order to constitute a system in equilibrium.

The statics thus constructed constitutes at the present day an important edifice to be henceforth classed amongst historical monuments. Some theorists even wish to go a step beyond. They have attempted to begin by the same means a more complete study of those systems whose state changes from one moment to another. This is, moreover, a study which is necessary to complete satisfactorily the study of equilibrium itself; for without it grave doubts would exist as to the conditions of stability, and it alone can give their true meaning to questions relating to displacements of equilibrium.

The problems with which we are thus confronted are singularly difficult. M. Duhem has given us many excellent examples of the fecundity of the method; but if thermodynamic statics may be considered definitely founded, it cannot be said that the general dynamics of systems, considered as the study of thermal movements and variations, are yet as solidly established.

§ 5. ATOMISM

It may appear singularly paradoxical that, in a chapter devoted to general views on the principles of physics, a few words should be introduced on the atomic theories of matter.

Very often, in fact, what is called the physics of principles is set in opposition to the hypotheses on the constitution of matter, particularly to atomic theories. I have already said that, abandoning the investigation of the unfathomable mystery of the constitution of the Universe, some physicists think they may find, in certain general principles, sufficient guides to conduct them across the physical world. But I have also said, in examining the history of those principles, that if they are to-day considered experimental truths, independent of all theories relating to matter, they have, in fact, nearly all been discovered by scholars who relied on molecular hypotheses: and the question suggests itself whether this is mere chance, or whether this chance may not be ordained by higher reasons.

In a very profound work which appeared a few years ago, entitled *Essai critique sur l'hypothese des atomes*, M. Hannequin, a philosopher who is also an erudite scholar, examined the part taken by atomism in the history of science. He notes that atomism and science were born, in Greece, of the same problem, and that in modern times the revival of the one was closely connected with that of the other. He shows, too, by very close analysis, that the atomic hypothesis is essential to the optics of Fresnel and of Cauchy; that it penetrates into the study of heat; and that, in its general features, it presided at the birth of modern chemistry and is linked with all its progress. He concludes that it is, in a manner, the soul of our knowledge of Nature, and that contemporary theories are on this point in accord with history: for these theories consecrate the preponderance of this hypothesis in the domain of science.

If M. Hannequin had not been prematurely cut off in the full expansion of his vigorous talent, he might have added another chapter to his excellent book. He would have witnessed a prodigious budding of atomistic ideas, accompanied, it is true, by wide modifications in the manner in which the atom is to be regarded, since the most recent theories make material atoms into centres constituted of atoms of electricity. On the other hand, he would have found in the bursting forth of these new doctrines one more proof in support of his idea that science is indissolubly bound to atomism.

From the philosophical point of view, M. Hannequin, examining the reasons which may have called these links into being, arrives at the idea that they necessarily proceed from the constitution of our knowledge, or, perhaps, from that of Nature itself. Moreover, this origin, double in appearance, is single at bottom. Our minds could not, in fact, detach and come out of themselves to grasp reality and the absolute in Nature. According to the idea of Descartes, it is the destiny of our minds only to take hold of and to understand that which proceeds from them.

Thus atomism, which is, perhaps, only an appearance containing even some contradictions, is yet a well-founded appearance, since it conforms to the laws of our minds; and this hypothesis is, in a way, necessary.

We may dispute the conclusions of M. Hannequin, but no one will refuse to recognise, as he does, that atomic theories occupy a preponderating part in the doctrines of physics; and the position which they have thus conquered gives them, in a way, the right of saying that they rest on a real principle. It is in order to recognise this right that several physicists—M. Langevin, for example—ask that atoms be promoted from the rank of hypotheses to that of principles. By this they mean that the atomistic ideas forced upon us by an almost obligatory induction based on very exact experiments, enable us to co-ordinate a considerable amount of facts, to construct a very general synthesis, and to foresee a great number of phenomena.

It is of moment, moreover, to thoroughly understand that atomism does not necessarily set up the hypothesis of centres of attraction acting at a distance, and it must not be confused with molecular physics, which has, on the other hand, undergone very serious checks. The molecular physics greatly in favour some fifty years ago leads to such complex representations and to solutions often so undetermined, that the most courageous are wearied with upholding it and it has fallen into some discredit. It rested on the fundamental principles of mechanics applied to molecular actions; and that was, no doubt, an extension legitimate enough, since mechanics is itself only an experimental science, and its principles, established for the movements of matter taken as a whole, should not be applied outside the domain which belongs to them. Atomism, in fact, tends more and more, in modern theories, to imitate the principle of the conservation of energy or that of entropy, to disengage itself from the artificial bonds which attached it to mechanics, and to put itself forward as an independent principle.

Atomistic ideas also have undergone evolution, and this slow evolution has been considerably quickened under the influence of modern discoveries. These reach back to the most remote antiquity, and to follow their development we should have to write the history of human thought which they have always accompanied since the time of Leucippus, Democritus, Epicurus, and Lucretius. The first observers who noticed that the volume of a body could be diminished by compression or cold, or augmented by heat, and who saw a soluble solid body mix completely with the water which dissolved it, must have been compelled to suppose that matter was not dispersed continuously throughout the space it seemed to occupy. They were thus brought to consider it discontinuous, and to admit that a substance having the same composition and the same properties in all its parts—in a word, perfectly homogeneous—ceases to present this homogeneity when considered within a sufficiently small volume.

Modern experimenters have succeeded by direct experiments in placing in evidence this heterogeneous character of matter when taken in small mass. Thus, for example, the superficial tension, which is constant for the same liquid at a given temperature, no longer has the same value when the thickness of the layer of liquid becomes extremely small. Newton noticed even in his time that a dark zone is seen to form on a soap bubble at the moment when it becomes so thin that it must burst. Professor Reinold and Sir Arthur Rücker have shown that this zone is no longer exactly spherical; and from this we must conclude that the superficial tension, constant for all thicknesses above a certain limit, commences to vary when the thickness falls below a critical value, which these authors estimate, on optical grounds, at about fifty millionths of a millimetre.

From experiments on capillarity, Prof. Quincke has obtained similar results with regard to layers of solids. But it is not only capillary properties which allow this characteristic to be revealed. All the properties of a body are modified when taken in small mass; M. Meslin proves this in a very ingenious way as regards optical properties, and Mr Vincent in respect of electric conductivity. M. Houllevigue, who, in a chapter of his excellent work, *Du Laboratoire à l'Usine*, has very clearly set forth the most interesting considerations on atomic hypotheses, has recently demonstrated that copper and silver cease to combine with iodine as soon as they are present in a thickness of less than thirty millionths of a millimetre. It is this same dimension likewise that is possessed, according to M. Wiener, by the smallest thicknesses it is possible to deposit on glass. These layers are so thin that they cannot be perceived, but their presence is revealed by a change in the properties of the light reflected by them.

Thus, below fifty to thirty millionths of a millimetre the properties of matter depend on its thickness. There are then, no doubt, only a few molecules to be met with, and it may be concluded, in consequence, that the discontinuous elements of bodies—that is, the molecules—have linear dimensions of the order of magnitude of the millionth of a millimetre. Considerations regarding more complex phenomena, for instance the phenomena of electricity by contact, and also the kinetic theory of gases, bring us to the same conclusion.

The idea of the discontinuity of matter forces itself upon us for many other reasons. All modern chemistry is founded on this principle; and laws like the law of multiple proportions, introduce an evident discontinuity to which we find analogies in the law of electrolysis. The elements of bodies we are thus brought to regard might, as regards solids at all events, be considered as immobile; but this immobility could not explain the phenomena of heat, and, as it is entirely inadmissible for gases, it seems very improbable it can absolutely occur in any state. We are thus led to suppose that these elements are animated by very complicated movements, each one proceeding in closed trajectories in which the least variations of temperature or pressure cause modifications.

The atomistic hypothesis shows itself remarkably fecund in the study of phenomena produced in gases, and here the mutual independence of the particles renders the question relatively more simple and, perhaps, allows the principles of mechanics to be more certainly extended to the movements of molecules.

The kinetic theory of gases can point to unquestioned successes; and the idea of Daniel Bernouilli, who, as early as 1738, considered a gaseous mass to be formed of a considerable number of molecules animated by rapid movements of translation, has been put into a form precise enough for mathematical analysis, and we have thus found ourselves in a position to construct a really solid foundation. It will be at once conceived, on this hypothesis, that pressure is the resultant of the shocks of the molecules against the walls of the containing vessel, and we at once come to the demonstration that the law of Mariotte is a natural consequence of this origin of pressure; since, if the volume occupied by a certain number of molecules is doubled, the number of shocks per second on each square centimetre of the walls becomes half as much. But if we attempt to carry this further, we find ourselves in presence of a serious difficulty. It is impossible to mentally follow every one of the many individual molecules which compose even a very limited mass of gas. The path followed by this molecule may be every instant modified by the chance of running against another, or by a shock which may make it rebound in another direction.

The difficulty would be insoluble if chance had not laws of its own. It was Maxwell who first thought of introducing into the kinetic theory the calculation of probabilities. Willard Gibbs and Boltzmann later on developed this idea, and have founded a statistical method which does not, perhaps, give absolute certainty, but which is certainly most interesting and curious.

Molecules are grouped in such a way that those belonging to the same group may be considered as having the same state of movement; then an examination is made of the number of molecules in each group, and what are the changes in this number from one moment to another. It is thus often possible to determine the part which the different groups have in the total properties of the system and in the phenomena which may occur.

Such a method, analogous to the one employed by statisticians for following the social phenomena in a population, is all the more legitimate the greater the number of individuals counted in the averages; now, the number of molecules contained in a limited space—for example, in a centimetre cube taken in normal conditions—is such that no population could ever attain so high a figure. All considerations, those we have indicated as well as others which might be invoked (for example, the recent researches of M. Spring on the limit of visibility of fluorescence), give this result:—that there are, in this space, some twenty thousand millions of molecules. Each of these must receive in the space of a millimetre about ten thousand shocks, and be ten thousand times thrust out of its course. The free path of a molecule is then very small, but it can be singularly augmented by diminishing the number of them. Tait and Dewar have calculated that, in a good modern vacuum, the length of the free path of the remaining molecules not taken away by the air-pump easily reaches a few centimetres.

By developing this theory, we come to consider that, for a given temperature, every molecule (and even every individual particle, atom, or ion) which takes part in the movement has, on the average, the same kinetic energy in every body, and that this energy is proportional to the absolute temperature; so that it is represented by this temperature multiplied by a constant quantity which is a universal constant.

This result is not an hypothesis but a very great probability. This probability increases when it is noted that the same value for the constant is met with in the study of very varied phenomena; for example, in certain theories on radiation. Knowing the mass and energy of a molecule, it is easy to calculate its speed; and we find that the average speed is about 400 metres per second for carbonic anhydride, 500 for nitrogen, and 1850 for hydrogen at 0° C. and at ordinary pressure. I shall have occasion, later on, to speak of much more considerable speeds than these as animating other particles.

The kinetic theory has permitted the diffusion of gases to be explained, and the divers circumstances of the phenomenon to be calculated. It has allowed us to show, as M. Brillouin has done, that the coefficient of diffusion of two gases does not depend on the proportion of the gases in the mixture; it gives a very striking image of the phenomena of viscosity and conductivity; and it leads us to think that the coefficients of friction and of conductivity are independent of the density; while all these previsions have been verified by experiment. It has also invaded optics; and by relying on the principle of Doppler, Professor Michelson has succeeded in obtaining from it an explanation of the length presented by the spectral rays of even the most rarefied gases.

But however interesting are these results, they would not have sufficed to overcome the repugnance of certain physicists for speculations which, an imposing mathematical baggage notwithstanding, seemed to them too hypothetical. The theory, moreover, stopped at the molecule, and appeared to suggest no idea which could lead to the discovery of the key to the phenomena where molecules exercise a mutual influence on each other. The kinetic hypothesis, therefore, remained in some disfavour with a great number of persons, particularly in France, until the last few years, when all the recent discoveries of the conductivity of gases and of the new radiations came to procure for it a new and luxuriant efflorescence. It may be said that the atomistic synthesis, but yesterday so decried, is to-day triumphant.

The elements which enter into the earlier kinetic theory, and which, to avoid confusion, should be always designated by the name of molecules, were not, truth to say, in the eyes of the chemists, the final term of the divisibility of matter. It is well known that, to them, except in certain particular bodies like the vapour of mercury and argon, the molecule comprises several atoms, and that, in compound bodies, the number of these atoms may even be fairly considerable. But physicists rarely needed to have recourse to the consideration of these atoms. They spoke of them to explain certain particularities of the propagation of sound, and to enunciate laws relating to specific heats; but, in general, they stopped at the consideration of the molecule.

The present theories carry the division much further. I shall not dwell now on these theories, since, in order to thoroughly understand them, many other facts must be examined. But to avoid all confusion, it remains understood that, contrary, no doubt, to etymology, but in conformity with present custom, I shall continue in what follows to call atoms those particles of matter which have till now been spoken of; these atoms being themselves, according to modern views, singularly complex edifices formed of elements, of which we shall have occasion to indicate the nature later.

CHAPTER IV
THE VARIOUS STATES OF MATTER
§ 1. THE STATICS OF FLUIDS

The division of bodies into gaseous, liquid, and solid, and the distinction established for the same substance between the three states, retain a great importance for the applications and usages of daily life, but have long since lost their absolute value from the scientific point of view.

So far as concerns the liquid and gaseous states particularly, the already antiquated researches of Andrews confirmed the ideas of Cagniard de la Tour and established the continuity of the two states. A group of physical studies has thus been constituted on what may be called the statics of fluids, in which we examine the relations existing between the pressure, the volume, and the temperature of bodies, and in which are comprised, under the term fluid, gases as well as liquids.

These researches deserve attention by their interest and the generality of the results to which they have led. They also give a remarkable example of the happy effects which may be obtained by the combined employment of the various methods of investigation used in exploring the domain of nature. Thermodynamics has, in fact, allowed us to obtain numerical relations between the various coefficients, and atomic hypotheses have led to the establishment of one capital relation, the characteristic equation of fluids; while, on the other hand, experiment in which the progress made in the art of measurement has been utilized, has furnished the most valuable information on all the laws of compressibility and dilatation.

The classical work of Andrews was not very wide. Andrews did not go much beyond pressures close to the normal and ordinary temperatures. Of late years several very interesting and peculiar cases have been examined by MM. Cailletet, Mathias, Batelli, Leduc, P. Chappuis, and other physicists. Sir W. Ramsay and Mr S. Young have made known the isothermal diagrams[6] of a certain number of liquid bodies at the ordinary temperature. They have thus been able, while keeping to somewhat restricted limits of temperature and pressure, to touch upon the most important questions, since they found themselves in the region of the saturation curve and of the critical point.

But the most complete and systematic body of researches is due to M. Amagat, who undertook the study of a certain number of bodies, some liquid and some gaseous, extending the scope of his experiments so as to embrace the different phases of the phenomena and to compare together, not only the results relating to the same bodies, but also those concerning different bodies which happen to be in the same conditions of temperature and pressure, but in very different conditions as regards their critical points.

From the experimental point of view, M. Amagat has been able, with extreme skill, to conquer the most serious difficulties. He has managed to measure with precision pressures amounting to 3000 atmospheres, and also the very small volumes then occupied by the fluid mass under consideration. This last measurement, which necessitates numerous corrections, is the most delicate part of the operation. These researches have dealt with a certain number of different bodies. Those relating to carbonic acid and ethylene take in the critical point. Others, on hydrogen and nitrogen, for instance, are very extended. Others, again, such as the study of the compressibility of water, have a special interest, on account of the peculiar properties of this substance. M. Amagat, by a very concise discussion of the experiments, has also been able to definitely establish the laws of compressibility and dilatation of fluids under constant pressure, and to determine the value of the various coefficients as well as their variations. It ought to be possible to condense all these results into a single formula representing the volume, the temperature, and the pressure. Rankin and, subsequently, Recknagel, and then Hirn, formerly proposed formulas of that kind; but the most famous, the one which first appeared to contain in a satisfactory manner all the facts which experiments brought to light and led to the production of many others, was the celebrated equation of Van der Waals.

Professor Van der Waals arrived at this relation by relying upon considerations derived from the kinetic theory of gases. If we keep to the simple idea at the bottom of this theory, we at once demonstrate that the gas ought to obey the laws of Mariotte and of Gay-Lussac, so that the characteristic equation would be obtained by the statement that the product of the number which is the measure of the volume by that which is the measure of the pressure is equal to a constant coefficient multiplied by the degree of the absolute temperature. But to get at this result we neglect two important factors.

We do not take into account, in fact, the attraction which the molecules must exercise on each other. Now, this attraction, which is never absolutely non-existent, may become considerable when the molecules are drawn closer together; that is to say, when the compressed gaseous mass occupies a more and more restricted volume. On the other hand, we assimilate the molecules, as a first approximation, to material points without dimensions; in the evaluation of the path traversed by each molecule no notice is taken of the fact that, at the moment of the shock, their centres of gravity are still separated by a distance equal to twice the radius of the molecule.

M. Van der Waals has sought out the modifications which must be introduced into the simple characteristic equation to bring it nearer to reality. He extends to the case of gases the considerations by which Laplace, in his famous theory of capillarity, reduced the effect of the molecular attraction to a perpendicular pressure exercised on the surface of a liquid. This leads him to add to the external pressure, that due to the reciprocal attractions of the gaseous particles. On the other hand, when we attribute finite dimensions to these particles, we must give a higher value to the number of shocks produced in a given time, since the effect of these dimensions is to diminish the mean path they traverse in the time which elapses between two consecutive shocks.

The calculation thus pursued leads to our adding to the pressure in the simple equation a term which is designated the internal pressure, and which is the quotient of a constant by the square of the volume; also to our deducting from the volume a constant which is the quadruple of the total and invariable volume which the gaseous molecules would occupy did they touch one another.

The experiments fit in fairly well with the formula of Van der Waals, but considerable discrepancies occur when we extend its limits, particularly when the pressures throughout a rather wider interval are considered; so that other and rather more complex formulas, on which there is no advantage in dwelling, have been proposed, and, in certain cases, better represent the facts.

But the most remarkable result of M. Van der Waals' calculations is the discovery of corresponding states. For a long time physicists spoke of bodies taken in a comparable state. Dalton, for example, pointed out that liquids have vapour-pressures equal to the temperatures equally distant from their boiling-point; but that if, in this particular property, liquids were comparable under these conditions of temperature, as regards other properties the parallelism was no longer to be verified. No general rule was found until M. Van der Waals first enunciated a primary law, viz., that if the pressure, the volume, and the temperature are estimated by taking as units the critical quantities, the constants special to each body disappear in the characteristic equation, which thus becomes the same for all fluids.

The words corresponding states thus take a perfectly precise signification. Corresponding states are those for which the numerical values of the pressure, volume, and temperature, expressed by taking as units the values corresponding to the critical point, are equal; and, in corresponding states any two fluids have exactly the same properties.

M. Natanson, and subsequently P. Curie and M. Meslin, have shown by various considerations that the same result may be arrived at by choosing units which correspond to any corresponding states; it has also been shown that the theorem of corresponding states in no way implies the exactitude of Van der Waals' formula. In reality, this is simply due to the fact that the characteristic equation only contains three constants.

The philosophical importance and the practical interest of the discovery nevertheless remain considerable. As was to be expected, numbers of experimenters have sought whether these consequences are duly verified in reality. M. Amagat, particularly, has made use for this purpose of a most original and simple method. He remarks that, in all its generality, the law may be translated thus: If the isothermal diagrams of two substances be drawn to the same scale, taking as unit of volume and of pressure the values of the critical constants, the two diagrams should coincide; that is to say, their superposition should present the aspect of one diagram appertaining to a single substance. Further, if we possess the diagrams of two bodies drawn to any scales and referable to any units whatever, as the changes of units mean changes in the scale of the axes, we ought to make one of the diagrams similar to the other by lengthening or shortening it in the direction of one of the axes. M. Amagat then photographs two isothermal diagrams, leaving one fixed, but arranging the other so that it may be free to turn round each axis of the co-ordinates; and by projecting, by means of a magic lantern, the second on the first, he arrives in certain cases at an almost complete coincidence.

This mechanical means of proof thus dispenses with laborious calculations, but its sensibility is unequally distributed over the different regions of the diagram. M. Raveau has pointed out an equally simple way of verifying the law, by remarking that if the logarithms of the pressure and volume are taken as co-ordinates, the co-ordinates of two corresponding points differ by two constant quantities, and the corresponding curves are identical.

From these comparisons, and from other important researches, among which should be particularly mentioned those of Mr S. Young and M. Mathias, it results that the laws of corresponding states have not, unfortunately, the degree of generality which we at first attributed to them, but that they are satisfactory when applied to certain groups of bodies.[7]

If in the study of the statics of a simple fluid the experimental results are already complex, we ought to expect much greater difficulties when we come to deal with mixtures; still the problem has been approached, and many points are already cleared up.

Mixed fluids may first of all be regarded as composed of a large number of invariable particles. In this particularly simple case M. Van der Waals has established a characteristic equation of the mixtures which is founded on mechanical considerations. Various verifications of this formula have been effected, and it has, in particular, been the object of very important remarks by M. Daniel Berthelot.

It is interesting to note that thermodynamics seems powerless to determine this equation, for it does not trouble itself about the nature of the bodies obedient to its laws; but, on the other hand, it intervenes to determine the properties of coexisting phases. If we examine the conditions of equilibrium of a mixture which is not subjected to external forces, it will be demonstrated that the distribution must come back to a juxtaposition of homogeneous phases; in a given volume, matter ought so to arrange itself that the total sum of free energy has a minimum value. Thus, in order to elucidate all questions relating to the number and qualities of the phases into which the substance divides itself, we are led to regard the geometrical surface which for a given temperature represents the free energy.

I am unable to enter here into the detail of the questions connected with the theories of Gibbs, which have been the object of numerous theoretical studies, and also of a series, ever more and more abundant, of experimental researches. M. Duhem, in particular, has published, on the subject, memoirs of the highest importance, and a great number of experimenters, mostly scholars working in the physical laboratory of Leyden under the guidance of the Director, Mr Kamerlingh Onnes, have endeavoured to verify the anticipations of the theory.

We are a little less advanced as regards abnormal substances; that is to say, those composed of molecules, partly simple and partly complex, and either dissociated or associated. These cases must naturally be governed by very complex laws. Recent researches by MM. Van der Waals, Alexeif, Rothmund, Künen, Lehfeld, etc., throw, however, some light on the question.

The daily more numerous applications of the laws of corresponding states have rendered highly important the determination of the critical constants which permit these states to be defined. In the case of homogeneous bodies the critical elements have a simple, clear, and precise sense; the critical temperature is that of the single isothermal line which presents a point of inflexion at a horizontal tangent; the critical pressure and the critical volume are the two co-ordinates of this point of inflexion.

The three critical constants may be determined, as Mr S. Young and M. Amagat have shown, by a direct method based on the consideration of the saturated states. Results, perhaps more precise, may also be obtained if one keeps to two constants or even to a single one—temperature, for example— by employing various special methods. Many others, MM. Cailletet and Colardeau, M. Young, M.J. Chappuis, etc., have proceeded thus.

The case of mixtures is much more complicated. A binary mixture has a critical space instead of a critical point. This space is comprised between two extreme temperatures, the lower corresponding to what is called the folding point, the higher to that which we call the point of contact of the mixture. Between these two temperatures an isothermal compression yields a quantity of liquid which increases, then reaches a maximum, diminishes, and disappears. This is the phenomenon of retrograde condensation. We may say that the properties of the critical point of a homogeneous substance are, in a way, divided, when it is a question of a binary mixture, between the two points mentioned.

Calculation has enabled M. Van der Waals, by the application of his kinetic theories, and M. Duhem, by means of thermodynamics, to foresee most of the results which have since been verified by experiment. All these facts have been admirably set forth and systematically co-ordinated by M. Mathias, who, by his own researches, moreover, has made contributions of the highest value to the study of questions regarding the continuity of the liquid and gaseous states.

The further knowledge of critical elements has allowed the laws of corresponding states to be more closely examined in the case of homogeneous substances. It has shown that, as I have already said, bodies must be arranged in groups, and this fact clearly proves that the properties of a given fluid are not determined by its critical constants alone, and that it is necessary to add to them some other specific parameters; M. Mathias and M. D. Berthelot have indicated some which seem to play a considerable part.

It results also from this that the characteristic equation of a fluid cannot yet be considered perfectly known. Neither the equation of Van der Waals nor the more complicated formulas which have been proposed by various authors are in perfect conformity with reality. We may think that researches of this kind will only be successful if attention is concentrated, not only on the phenomena of compressibility and dilatation, but also on the calorimetric properties of bodies. Thermodynamics indeed establishes relations between those properties and other constants, but does not allow everything to be foreseen.

Several physicists have effected very interesting calorimetric measurements, either, like M. Perot, in order to verify Clapeyron's formula regarding the heat of vaporization, or to ascertain the values of specific heats and their variations when the temperature or the pressure happens to change. M. Mathias has even succeeded in completely determining the specific heats of liquefied gases and of their saturated vapours, as well as the heat of internal and external vaporization.

§ 2. THE LIQUEFACTION OF GASES, AND THE PROPERTIES OF BODIES AT A LOW TEMPERATURE

The scientific advantages of all these researches have been great, and, as nearly always happens, the practical consequences derived from them have also been most important. It is owing to the more complete knowledge of the general properties of fluids that immense progress has been made these last few years in the methods of liquefying gases.

From a theoretical point of view the new processes of liquefaction can be classed in two categories. Linde's machine and those resembling it utilize, as is known, expansion without any notable production of external work. This expansion, nevertheless, causes a fall in the temperature, because the gas in the experiment is not a perfect gas, and, by an ingenious process, the refrigerations produced are made cumulative.

Several physicists have proposed to employ a method whereby liquefaction should be obtained by expansion with recuperable external work. This method, proposed as long ago as 1860 by Siemens, would offer considerable advantages. Theoretically, the liquefaction would be more rapid, and obtained much more economically; but unfortunately in the experiment serious obstacles are met with, especially from the difficulty of obtaining a suitable lubricant under intense cold for those parts of the machine which have to be in movement if the apparatus is to work.

M. Claude has recently made great progress on this point by the use, during the running of the machine, of the ether of petrol, which is uncongealable, and a good lubricant for the moving parts. When once the desired region of cold is reached, air itself is used, which moistens the metals but does not completely avoid friction; so that the results would have remained only middling, had not this ingenious physicist devised a new improvement which has some analogy with superheating of steam in steam engines. He slightly varies the initial temperature of the compressed air on the verge of liquefaction so as to avoid a zone of deep perturbations in the properties of fluids, which would make the work of expansion very feeble and the cold produced consequently slight. This improvement, simple as it is in appearance, presents several other advantages which immediately treble the output.

The special object of M. Claude was to obtain oxygen in a practical manner by the actual distillation of liquid air. Since nitrogen boils at -194° and oxygen at -180.5° C., if liquid air be evaporated, the nitrogen escapes, especially at the commencement of the evaporation, while the oxygen concentrates in the residual liquid, which finally consists of pure oxygen, while at the same time the temperature rises to the boiling-point (-180.5° C.) of oxygen. But liquid air is costly, and if one were content to evaporate it for the purpose of collecting a part of the oxygen in the residuum, the process would have a very poor result from the commercial point of view. As early as 1892, Mr Parkinson thought of improving the output by recovering the cold produced by liquid air during its evaporation; but an incorrect idea, which seems to have resulted from certain experiments of Dewar—the idea that the phenomenon of the liquefaction of air would not be, owing to certain peculiarities, the exact converse of that of vaporization—led to the employment of very imperfect apparatus. M. Claude, however, by making use of a method which he calls the reversal [8] method, obtains a complete rectification in a remarkably simple manner and under extremely advantageous economic conditions. Apparatus, of surprisingly reduced dimensions but of great efficiency, is now in daily work, which easily enables more than a thousand cubic metres of oxygen to be obtained at the rate, per horse-power, of more than a cubic metre per hour.

It is in England, thanks to the skill of Sir James Dewar and his pupils— thanks also, it must be said, to the generosity of the Royal Institution, which has devoted considerable sums to these costly experiments—that the most numerous and systematic researches have been effected on the production of intense cold. I shall here note only the more important results, especially those relating to the properties of bodies at low temperatures.

Their electrical properties, in particular, undergo some interesting modifications. The order which metals assume in point of conductivity is no longer the same as at ordinary temperatures. Thus at -200° C. copper is a better conductor than silver. The resistance diminishes with the temperature, and, down to about -200°, this diminution is almost linear, and it would seem that the resistance tends towards zero when the temperature approaches the absolute zero. But, after -200°, the pattern of the curves changes, and it is easy to foresee that at absolute zero the resistivities of all metals would still have, contrary to what was formerly supposed, a notable value. Solidified electrolytes which, at temperatures far below their fusion point, still retain a very appreciable conductivity, become, on the contrary, perfect insulators at low temperatures. Their dielectric constants assume relatively high values. MM. Curie and Compan, who have studied this question from their own point of view, have noted, moreover, that the specific inductive capacity changes considerably with the temperature.

In the same way, magnetic properties have been studied. A very interesting result is that found in oxygen: the magnetic susceptibility of this body increases at the moment of liquefaction. Nevertheless, this increase, which is enormous (since the susceptibility becomes sixteen hundred times greater than it was at first), if we take it in connection with equal volumes, is much less considerable if taken in equal masses. It must be concluded from this fact that the magnetic properties apparently do not belong to the molecules themselves, but depend on their state of aggregation.

The mechanical properties of bodies also undergo important modifications. In general, their cohesion is greatly increased, and the dilatation produced by slight changes of temperature is considerable. Sir James Dewar has effected careful measurements of the dilatation of certain bodies at low temperatures: for example, of ice. Changes in colour occur, and vermilion and iodide of mercury pass into pale orange. Phosphorescence becomes more intense, and most bodies of complex structure—milk, eggs, feathers, cotton, and flowers—become phosphorescent. The same is the case with certain simple bodies, such as oxygen, which is transformed into ozone and emits a white light in the process.

Chemical affinity is almost put an end to; phosphorus and potassium remain inert in liquid oxygen. It should, however, be noted, and this remark has doubtless some interest for the theories of photographic action, that photographic substances retain, even at the temperature of liquid hydrogen, a very considerable part of their sensitiveness to light.

Sir James Dewar has made some important applications of low temperatures in chemical analysis; he also utilizes them to create a vacuum. His researches have, in fact, proved that the pressure of air congealed by liquid hydrogen cannot exceed the millionth of an atmosphere. We have, then, in this process, an original and rapid means of creating an excellent vacuum in apparatus of very different kinds—a means which, in certain cases, may be particularly convenient.[9]

Thanks to these studies, a considerable field has been opened up for biological research, but in this, which is not our subject, I shall notice one point only. It has been proved that vital germs—bacteria, for example—may be kept for seven days at -190°C. without their vitality being modified. Phosphorescent organisms cease, it is true, to shine at the temperature of liquid air, but this fact is simply due to the oxidations and other chemical reactions which keep up the phosphorescence being then suspended, for phosphorescent activity reappears so soon as the temperature is again sufficiently raised. An important conclusion has been drawn from these experiments which affects cosmogonical theories: since the cold of space could not kill the germs of life, it is in no way absurd to suppose that, under proper conditions, a germ may be transmitted from one planet to another.

Among the discoveries made with the new processes, the one which most strikingly interested public attention is that of new gases in the atmosphere. We know how Sir William Ramsay and Dr. Travers first observed by means of the spectroscope the characteristics of the *companions* of argon in the least volatile part of the atmosphere. Sir James Dewar on the one hand, and Sir William Ramsay on the other, subsequently separated in addition to argon and helium, crypton, xenon, and neon. The process employed consists essentially in first solidifying the least volatile part of the air and then causing it to evaporate with extreme slowness. A tube with electrodes enables the spectrum of the gas in process of distillation to be observed. In this manner, the spectra of the various gases may be seen following one another in the inverse order of their volatility. All these gases are monoatomic, like mercury; that is to say, they are in the most simple state, they possess no internal molecular energy (unless it is that which heat is capable of supplying), and they even seem to have no chemical energy. Everything leads to the belief that they show the existence on the earth of an earlier state of things now vanished. It may be supposed, for instance, that helium and neon, of which the molecular mass is very slight, were formerly more abundant on our planet; but at an epoch when the temperature of the globe was higher, the very speed of their molecules may have reached a considerable value, exceeding, for instance, eleven kilometres per second, which suffices to explain why they should have left our atmosphere. Crypton and neon, which have a density four times greater than oxygen, may, on the contrary, have partly disappeared by solution at the bottom of the sea, where it is not absurd to suppose that considerable quantities would be found liquefied at great depths. [10]

It is probable, moreover, that the higher regions of the atmosphere are not composed of the same air as that around us. Sir James Dewar points out that Dalton's law demands that every gas composing the atmosphere should have, at all heights and temperatures, the same pressure as if it were alone, the pressure decreasing the less quickly, all things being equal, as its density becomes less. It results from this that the temperature becoming gradually lower as we rise in the atmosphere, at a certain altitude there can no longer remain any traces of oxygen or nitrogen, which no doubt liquefy, and the atmosphere must be almost exclusively composed of the most volatile gases, including hydrogen, which M.A. Gautier has, like Lord Rayleigh and Sir William Ramsay, proved to exist in the air. The spectrum of the *Aurora borealis*, in which are found the lines of those parts of the atmosphere which cannot be liquefied in liquid hydrogen, together with the lines of argon, crypton, and xenon, is quite in conformity with this point of view. It is, however, singular that it should be the spectrum of crypton, that is to say, of the heaviest gas of the group, which appears most clearly in the upper regions of the atmosphere.

Among the gases most difficult to liquefy, hydrogen has been the object of particular research and of really quantitative experiments. Its properties in a liquid state are now very clearly known. Its boiling-point, measured with a helium thermometer which has been compared with thermometers of oxygen and hydrogen, is -252°; its critical temperature is -241° C.; its critical pressure, 15 atmospheres. It is four times lighter than water, it does not present any absorption spectrum, and its specific heat is the greatest known. It is not a conductor of electricity. Solidified at 15° absolute, it is far from reminding one by its aspect of a metal; it rather resembles a piece of perfectly pure ice, and Dr Travers attributes to it a crystalline structure. The last gas which has resisted liquefaction, helium, has recently been obtained in a liquid state; it appears to have its boiling-point in the neighbourhood of 6° absolute. [11]

§ 3. SOLIDS AND LIQUIDS

The interest of the results to which the researches on the continuity between the liquid and the gaseous states have led is so great, that numbers of scholars have naturally been induced to inquire whether something analogous might not be found in the case of liquids and solids. We might think that a similar continuity ought to be there met with, that the universal character of the properties of matter forbade all real discontinuity between two different states, and that, in truth, the solid was a prolongation of the liquid state.

To discover whether this supposition is correct, it concerns us to compare the properties of liquids and solids. If we find that all properties are common to the two states we have the right to believe, even if they presented themselves in different degrees, that, by a continuous series of intermediary bodies, the two classes might yet be connected. If, on the other hand, we discover that there exists in these two classes some quality of a different nature, we must necessarily conclude that there is a discontinuity which nothing can remove.

The distinction established, from the point of view of daily custom, between solids and liquids, proceeds especially from the difficulty that we meet with in the one case, and the facility in the other, when we wish to change their form temporarily or permanently by the action of mechanical force. This distinction only corresponds, however, in reality, to a difference in the value of certain coefficients. It is impossible to discover by this means any absolute characteristic which establishes a separation between the two classes. Modern researches prove this clearly. It is not without use, in order to well understand them, to state precisely the meaning of a few terms generally rather loosely employed.

If a conjunction of forces acting on a homogeneous material mass happens to deform it without compressing or dilating it, two very distinct kinds of reactions may appear which oppose themselves to the effort exercised. During the time of deformation, and during that time only, the first make their influence felt. They depend essentially on the greater or less rapidity of the deformation, they cease with the movement, and could not, in any case, bring the body back to its pristine state of equilibrium. The existence of these reactions leads us to the idea of viscosity or internal friction.

The second kind of reactions are of a different nature. They continue to act when the deformation remains stationary, and, if the external forces happen to disappear, they are capable of causing the body to return to its initial form, provided a certain limit has not been exceeded. These last constitute rigidity.

At first sight a solid body appears to have a finite rigidity and an infinite viscosity; a liquid, on the contrary, presents a certain viscosity, but no rigidity. But if we examine the matter more closely, beginning either with the solids or with the liquids, we see this distinction vanish.

Tresca showed long ago that internal friction is not infinite in a solid; certain bodies can, so to speak, at once flow and be moulded. M.W. Spring has given many examples of such phenomena. On the other hand, viscosity in liquids is never non-existent; for were it so for water, for example, in the celebrated experiment effected by Joule for the determination of the mechanical equivalent of the caloric, the liquid borne along by the floats would slide without friction on the surrounding liquid, and the work done by movement would be the same whether the floats did or did not plunge into the liquid mass.

In certain cases observed long ago with what are called pasty bodies, this viscosity attains a value almost comparable to that observed by M. Spring in some solids. Nor does rigidity allow us to establish a barrier between the two states. Notwithstanding the extreme mobility of their particles, liquids contain, in fact, vestiges of the property which we formerly wished to consider the special characteristic of solids.

Maxwell before succeeded in rendering the existence of this rigidity very probable by examining the optical properties of a deformed layer of liquid. But a Russian physicist, M. Schwedoff, has gone further, and has been able by direct experiments to show that a sheath of liquid set between two solid cylinders tends, when one of the cylinders is subjected to a slight rotation, to return to its original position, and gives a measurable torsion to a thread upholding the cylinder. From the knowledge of this torsion the rigidity can be deduced. In the case of a solution containing 1/2 per cent. of gelatine, it is found that this rigidity, enormous compared with that of water, is still, however, one trillion eight hundred and forty billion times less than that of steel.

This figure, exact within a few billions, proves that the rigidity is very slight, but exists; and that suffices for a characteristic distinction to be founded on this property. In a general way, M. Spring has also established that we meet in solids, in a degree more or less marked, with the properties of liquids. When they are placed in suitable conditions of pressure and time, they flow through orifices, transmit pressure in all directions, diffuse and dissolve one into the other, and react chemically on each other. They may be soldered together by compression; by the same means alloys may be produced; and further, which seems to clearly prove that matter in a solid state is not deprived of all molecular mobility, it is possible to realise suitable limited reactions and equilibria between solid salts, and these equilibria obey the fundamental laws of thermodynamics.

Thus the definition of a solid cannot be drawn from its mechanical properties. It cannot be said, after what we have just seen, that solid bodies retain their form, nor that they have a limited elasticity, for M. Spring has made known a case where the elasticity of solids is without any limit.

It was thought that in the case of a different phenomenon—that of crystallization—we might arrive at a clear distinction, because here we should he dealing with a specific quality; and that crystallized bodies would be the true solids, amorphous bodies being at that time regarded as liquids viscous in the extreme.

But the studies of a German physicist, Professor 0. Lehmann, seem to prove that even this means is not infallible. Professor Lehmann has succeeded, in fact, in obtaining with certain organic compounds—oleate of potassium, for instance—under certain conditions some peculiar states to which he has given the name of semi-fluid and liquid crystals. These singular phenomena can only be observed and studied by means of a microscope, and the Carlsruhe Professor had to devise an ingenious apparatus which enabled him to bring the preparation at the required temperature on to the very plate of the microscope.

It is thus made evident that these bodies act on polarized light in the manner of a crystal. Those that M. Lehmann terms semi-liquid still present traces of polyhedric delimitation, but with the peaks and angles rounded by surface-tension, while the others tend to a strictly spherical form. The optical examination of the first-named bodies is very difficult, because appearances may be produced which are due to the phenomena of refraction and imitate those of polarization. For the other kind, which are often as mobile as water, the fact that they polarize light is absolutely unquestionable.

Unfortunately, all these liquids are turbid, and it may be objected that they are not homogeneous. This want of homogeneity may, according to M. Quincke, be due to the existence of particles suspended in a liquid in contact with another liquid miscible with it and enveloping it as might a membrane, and the phenomena of polarization would thus be quite naturally explained. [12]

M. Tamman is of opinion that it is more a question of an emulsion, and, on this hypothesis, the action on light would actually be that which has been observed. Various experimenters have endeavoured of recent years to elucidate this question. It cannot be considered absolutely settled, but these very curious experiments, pursued with great patience and remarkable ingenuity, allow us to think that there really exist certain intermediary forms between crystals and liquids in which bodies still retain a peculiar structure, and consequently act on light, but nevertheless possess considerable plasticity.

Let us note that the question of the continuity of the liquid and solid states is not quite the same as the question of knowing whether there exist bodies intermediate in all respects between the solids and liquids. These two problems are often wrongly confused. The gap between the two classes of bodies may be filled by certain substances with intermediate properties, such as pasty bodies and bodies liquid but still crystallized, because they have not yet completely lost their peculiar structure. Yet the transition is not necessarily established in a continuous fashion when we are dealing with the passage of one and the same determinate substance from the liquid to the solid form. We conceive that this change may take place by insensible degrees in the case of an amorphous body. But it seems hardly possible to consider the case of a crystal, in which molecular movements must be essentially regular, as a natural sequence to the case of the liquid where we are, on the contrary, in presence of an extremely disordered state of movement.

M. Taminan has demonstrated that amorphous solids may very well, in fact, be regarded as superposed liquids endowed with very great viscosity. But it is no longer the same thing when the solid is once in the crystallized state. There is then a solution of continuity of the various properties of the substance, and the two phases may co-exist.

We might presume also, by analogy with what happens with liquids and gases, that if we followed the curve of transformation of the crystalline into the liquid phase, we might arrive at a kind of critical point at which the discontinuity of their properties would vanish.

Professor Poynting, and after him Professor Planck and Professor Ostwald, supposed this to be the case, but more recently M. Tamman has shown that such a point does not exist, and that the region of stability of the crystallized state is limited on all sides. All along the curve of transformation the two states may exist in equilibrium, but we may assert that it is impossible to realize a continuous series of intermediaries between these two states. There will always be a more or less marked discontinuity in some of the properties.

In the course of his researches M. Tamman has been led to certain very important observations, and has met with fresh allotropic modifications in nearly all substances, which singularly complicate the question. In the case of water, for instance, he finds that ordinary ice transforms itself, under a given pressure, at the temperature of -80° C. into another crystalline variety which is denser than water.

The statics of solids under high pressure is as yet, therefore, hardly drafted, but it seems to promise results which will not be identical with those obtained for the statics of fluids, though it will present at least an equal interest.

§ 4. THE DEFORMATIONS OF SOLIDS

If the mechanical properties of the bodies intermediate between solids and liquids have only lately been the object of systematic studies, admittedly solid substances have been studied for a long time. Yet, notwithstanding the abundance of researches published on elasticity by theorists and experimenters, numerous questions with regard to them still remain in suspense.

We only propose to briefly indicate here a few problems recently examined, without going into the details of questions which belong more to the domain of mechanics than to that of pure physics.

The deformations produced in solid bodies by increasing efforts arrange themselves in two distinct periods. If the efforts are weak, the deformations produced are also very weak and disappear when the effort ceases. They are then termed elastic. If the efforts exceed a certain value, a part only of these deformations disappear, and a part are permanent.

The purity of the note emitted by a sound has been often invoked as a proof of the perfect isochronism of the oscillation, and, consequently, as a demonstration *a posteriori* of the correctness of the early law of Hoocke governing elastic deformations. This law has, however, during some years been frequently disputed. Certain mechanicians or physicists freely admit it to be incorrect, especially as regards extremely weak deformations. According to a theory in some favour, especially in Germany, *i.e.* the theory of Bach, the law which connects the elastic deformations with the efforts would be an exponential one. Recent experiments by Professors Kohlrausch and Gruncisen, executed under varied and precise conditions on brass, cast iron, slate, and wrought iron, do not appear to confirm Bach's law. Nothing, in point of fact, authorises the rejection of the law of Hoocke, which presents itself as the most natural and most simple approximation to reality.

The phenomena of permanent deformation are very complex, and it certainly seems that they cannot be explained by the older theories which insisted that the molecules only acted along the straight line which joined their centres. It becomes necessary, then, to construct more complete hypotheses, as the MM. Cosserat have done in some excellent memoirs, and we may then succeed in grouping together the facts resulting from new experiments. Among the experiments of which every theory must take account may be mentioned those by which Colonel Hartmann has placed in evidence the importance of the lines which are produced on the surface of metals when the limit of elasticity is exceeded.

It is to questions of the same order that the minute and patient researches of M. Bouasse have been directed. This physicist, as ingenious as he is profound, has pursued for several years experiments on the most delicate points relating to the theory of elasticity, and he has succeeded in defining with a precision not always attained even in the best esteemed works, the deformations to which a body must be subjected in order to obtain comparable experiments. With regard to the slight oscillations of torsion which he has specially studied, M. Bouasse arrives at the conclusion, in an acute discussion, that we hardly know anything more than was proclaimed a hundred years ago by Coulomb. We see, by this example, that admirable as is the progress accomplished in certain regions of physics, there still exist many over-neglected regions which remain in painful darkness. The skill shown by M. Bouasse authorises us to hope that, thanks to his researches, a strong light will some day illumine these unknown corners.

A particularly interesting chapter on elasticity is that relating to the study of crystals; and in the last few years it has been the object of remarkable researches on the part of M. Voigt. These researches have permitted a few controversial questions between theorists and experimenters to be solved: in particular, M. Voigt has verified the consequences of the calculations, taking care not to make, like Cauchy and Poisson, the hypothesis of central forces a mere function of distance, and has recognized a potential which depends on the relative orientation of the molecules. These considerations also apply to quasi-isotropic bodies which are, in fact, networks of crystals.

Certain occasional deformations which are produced and disappear slowly may be considered as intermediate between elastic and permanent deformations. Of these, the thermal deformation of glass which manifests itself by the displacement of the zero of a thermometer is an example. So also the modifications which the phenomena of magnetic hysteresis or the variations of resistivity have just demonstrated.

Many theorists have taken in hand these difficult questions. M. Brillouin endeavours to interpret these various phenomena by the molecular hypothesis. The attempt may seem bold, since these phenomena are, for the most part, essentially irreversible, and seem, consequently, not adaptable to mechanics. But M. Brillouin makes a point of showing that, under certain conditions, irreversible phenomena may be created between two material points, the actions of which depend solely on their distance; and he furnishes striking instances which appear to prove that a great number of irreversible physical and chemical phenomena may be ascribed to the existence of states of unstable equilibria.

M. Duhem has approached the problem from another side, and endeavours to bring it within the range of thermodynamics. Yet ordinary thermodynamics could not account for experimentally realizable states of equilibrium in the phenomena of viscosity and friction, since this science declares them to be impossible. M. Duhem, however, arrives at the idea that the establishment of the equations of thermodynamics presupposes, among other hypotheses, one which is entirely arbitrary, namely: that when the state of the system is given, external actions capable of maintaining it in that state are determined without ambiguity, by equations termed conditions of equilibrium of the system. If we reject this hypothesis, it will then be allowable to introduce into thermodynamics laws previously excluded, and it will be possible to construct, as M. Duhem has done, a much more comprehensive theory.

The ideas of M. Duhem have been illustrated by remarkable experimental work. M. Marchis, for example, guided by these ideas, has studied the permanent modifications produced in glass by an oscillation of temperature. These modifications, which may be called phenomena of the hysteresis of dilatation, may be followed in very appreciable fashion by means of a glass thermometer. The general results are quite in accord with the previsions of M. Duhem. M. Lenoble in researches on the traction of metallic wires, and M. Chevalier in experiments on the permanent variations of the electrical resistance of wires of an alloy of platinum and silver when submitted to periodical variations of temperature, have likewise afforded verifications of the theory propounded by M. Duhem.

In this theory, the representative system is considered dependent on the temperature of one or several other variables, such as, for example, a chemical variable. A similar idea has been developed in a very fine set of memoirs on nickel steel, by M. Ch. Ed. Guillaume. The eminent physicist, who, by his earlier researches, has greatly contributed to the light thrown on the analogous question of the displacement of the zero in thermometers, concludes, from fresh researches, that the residual phenomena are due to chemical variations, and that the return to the primary chemical state causes the variation to disappear. He applies his ideas not only to the phenomena presented by irreversible steels, but also to very different facts; for example, to phosphorescence, certain particularities of which may be interpreted in an analogous manner.

Nickel steels present the most curious properties, and I have already pointed out the paramount importance of one of them, hardly capable of perceptible dilatation, for its application to metrology and chronometry. [13] Others, also discovered by M. Guillaume in the course of studies conducted with rare success and remarkable ingenuity, may render great services, because it is possible to regulate, so to speak, at will their mechanical or magnetic properties.

The study of alloys in general is, moreover, one of those in which the introduction of the methods of physics has produced the greatest effects. By the microscopic examination of a polished surface or of one indented by a reagent, by the determination of the electromotive force of elements of which an alloy forms one of the poles, and by the measurement of the resistivities, the densities, and the differences of potential or contact, the most valuable indications as to their constitution are obtained. M. Le Chatelier, M. Charpy, M. Dumas, M. Osmond, in France; Sir W. Roberts Austen and Mr. Stansfield, in England, have given manifold examples of the fertility of these methods. The question, moreover, has had a new light thrown upon it by the application of the principles of thermodynamics and of the phase rule.

Alloys are generally known in the two states of solid and liquid. Fused alloys consist of one or several solutions of the component metals and of a certain number of definite combinations. Their composition may thus be very complex: but Gibbs' rule gives us at once important information on the point, since it indicates that there cannot exist, in general, more than two distinct solutions in an alloy of two metals.

Solid alloys may be classed like liquid ones. Two metals or more dissolve one into the other, and form a solid solution quite analogous to the liquid solution. But the study of these solid solutions is rendered singularly difficult by the fact that the equilibrium so rapidly reached in the case of liquids in this case takes days and, in certain cases, perhaps even centuries to become established.

CHAPTER V
SOLUTIONS AND ELECTROLYTIC DISSOCIATION

§ 1. SOLUTION

Vaporization and fusion are not the only means by which the physical state of a body may be changed without modifying its chemical constitution. From the most remote periods solution has also been known and studied, but only in the last twenty years have we obtained other than empirical information regarding this phenomenon.

It is natural to employ here also the methods which have allowed us to penetrate into the knowledge of other transformations. The problem of solution may be approached by way of thermodynamics and of the hypotheses of kinetics.

As long ago as 1858, Kirchhoff, by attributing to saline solutions—that is to say, to mixtures of water and a non-volatile liquid like sulphuric acid—the properties of internal energy, discovered a relation between the quantity of heat given out on the addition of a certain quantity of water to a solution and the variations to which condensation and temperature subject the vapour-tension of the solution. He calculated for this purpose the variations of energy which are produced when passing from one state to another by two different series of transformations; and, by comparing the two expressions thus obtained, he established a relation between the various elements of the phenomenon. But, for a long time afterwards, the question made little progress, because there seemed to be hardly any means of introducing into this study the second principle of thermodynamics. [14] It was the memoir of Gibbs which at last opened out this rich domain and enabled it to be rationally exploited. As early as 1886, M. Duhem showed that the theory of the thermodynamic potential furnished precise information on solutions or liquid mixtures. He thus discovered over again the famous law on the lowering of the congelation temperature of solvents which had just been established by M. Raoult after a long series of now classic researches.

In the minds of many persons, however, grave doubts persisted. Solution appeared to be an essentially irreversible phenomenon. It was therefore, in all strictness, impossible to calculate the entropy of a solution, and consequently to be certain of the value of the thermodynamic potential. The objection would be serious even to-day, and, in calculations, what is called the paradox of Gibbs would be an obstacle.

We should not hesitate, however, to apply the Phase Law to solutions, and this law already gives us the key to a certain number of facts. It puts in evidence, for example, the part played by the eutectic point—that is to say, the point at which (to keep to the simple case in which we have to do with two bodies only, the solvent and the solute) the solution is in equilibrium at once with the two possible solids, the dissolved body and the solvent solidified. The knowledge of this point explains the properties of refrigerating mixtures, and it is also one of the most useful for the theory of alloys. The scruples of physicists ought to have been removed on the memorable occasion when Professor Van t'Hoff demonstrated that solution can operate reversibly by reason of the phenomena of osmosis. But the experiment can only succeed in very rare cases; and, on the other hand, Professor Van t'Hoff was naturally led to another very bold conception. He regarded the molecule of the dissolved body as a gaseous one, and assimilated solution, not as had hitherto been the rule, to fusion, but to a kind of vaporization. Naturally his ideas were not immediately accepted by the scholars most closely identified with the classic tradition. It may perhaps not be without use to examine here the principles of Professor Van t'Hoff's theory.

§ 2. OSMOSIS

Osmosis, or diffusion through a septum, is a phenomenon which has been known for some time. The discovery of it is attributed to the Abbé Nollet, who is supposed to have observed it in 1748, during some "researches on liquids in ebullition." A classic experiment by Dutrochet, effected about 1830, makes this phenomenon clear. Into pure water is plunged the lower part of a vertical tube containing pure alcohol, open at the top and closed at the bottom by a membrane, such as a pig's bladder, without any visible perforation. In a very short time it will be found, by means of an areometer for instance, that the water outside contains alcohol, while the alcohol of the tube, pure at first, is now diluted. Two currents have therefore passed through the membrane, one of water from the outside to the inside, and one of alcohol in the converse direction. It is also noted that a difference in the levels has occurred, and that the liquid in the tube now rises to a considerable height. It must therefore be admitted that the flow of the water has been more rapid than that of the alcohol. At the commencement, the water must have penetrated into the tube much more rapidly than the alcohol left it. Hence the difference in the levels, and, consequently, a difference of pressure on the two faces of the membrane. This difference goes on increasing, reaches a maximum, then diminishes, and vanishes when the diffusion is complete, final equilibrium being then attained.

The phenomenon is evidently connected with diffusion. If water is very carefully poured on to alcohol, the two layers, separate at first, mingle by degrees till a homogeneous substance is obtained. The bladder seems not to have prevented this diffusion from taking place, but it seems to have shown itself more permeable to water than to alcohol. May it not therefore be supposed that there must exist dividing walls in which this difference of permeability becomes greater and greater, which would be permeable to the solvent and absolutely impermeable to the solute? If this be so, the phenomena of these *semi-permeable* walls, as they are termed, can be observed in particularly simple conditions.

The answer to this question has been furnished by biologists, at which we cannot be surprised. The phenomena of osmosis are naturally of the first importance in the action of organisms, and for a long time have attracted the attention of naturalists. De Vries imagined that the contractions noticed in the protoplasm of cells placed in saline solutions were due to a phenomenon of osmosis, and, upon examining more closely certain peculiarities of cell life, various scholars have demonstrated that living cells are enclosed in membranes permeable to certain substances and entirely impermeable to others. It was interesting to try to reproduce artificially semi-permeable walls analogous to those thus met with in nature; [15] and Traube and Pfeffer seem to have succeeded in one particular case. Traube has pointed out that the very delicate membrane of ferrocyanide of potassium which is obtained with some difficulty by exposing it to the reaction of sulphate of copper, is permeable to water, but will not permit the passage of the majority of salts. Pfeffer, by producing these walls in the interstices of a porous porcelain, has succeeded in giving them sufficient rigidity to allow measurements to be made. It must be allowed that, unfortunately, no physicist or chemist has been as lucky as these two botanists; and the attempts to reproduce semi-permeable walls completely answering to the definition, have never given but mediocre results. If, however, the experimental difficulty has not been overcome in an entirely satisfactory manner, it at least appears very probable that such walls may nevertheless exist. [16]

Nevertheless, in the case of gases, there exists an excellent example of a semi-permeable wall, and a partition of platinum brought to a higher than red heat is, as shown by M. Villard in some ingenious experiments, completely impermeable to air, and very permeable, on the contrary, to hydrogen. It can also be experimentally demonstrated that on taking two recipients separated by such a partition, and both containing nitrogen mixed with varying proportions of hydrogen, the last-named gas will pass through the partition in such a way that the concentration—that is to say, the mass of gas per unit of volume—will become the same on both sides. Only then will equilibrium be established; and, at that moment, an excess of pressure will naturally be produced in that recipient which, at the commencement, contained the gas with the smallest quantity of hydrogen.

This experiment enables us to anticipate what will happen in a liquid medium with semi-permeable partitions. Between two recipients, one containing pure water, the other, say, water with sugar in solution, separated by one of these partitions, there will be produced merely a movement of the pure towards the sugared water, and following this, an increase of pressure on the side of the last. But this increase will not be without limits. At a certain moment the pressure will cease to increase and will remain at a fixed value which now has a given direction. This is the osmotic pressure.

Pfeffer demonstrated that, for the same substance, the osmotic pressure is proportional to the concentration, and consequently in inverse ratio to the volume occupied by a similar mass of the solute. He gave figures from which it was easy, as Professor Van t'Hoff found, to draw the conclusion that, in a constant volume, the osmotic pressure is proportional to the absolute temperature. De Vries, moreover, by his remarks on living cells, extended the results which Pfeffer had applied to one case only—that is, to the one that he had been able to examine experimentally.

Such are the essential facts of osmosis. We may seek to interpret them and to thoroughly examine the mechanism of the phenomenon; but it must be acknowledged that as regards this point, physicists are not entirely in accord. In the opinion of Professor Nernst, the permeability of semi-permeable membranes is simply due to differences of solubility in one of the substances of the membrane itself. Other physicists think it attributable, either to the difference in the dimensions of the molecules, of which some might pass through the pores of the membrane and others be stopped by their relative size, or to these molecules' greater or less mobility. For others, again, it is the capillary phenomena which here act a preponderating part.

This last idea is already an old one: Jager, More, and Professor Traube have all endeavoured to show that the direction and speed of osmosis are determined by differences in the surface-tensions; and recent experiments, especially those of Batelli, seem to prove that osmosis establishes itself in the way which best equalizes the surface-tensions of the liquids on both sides of the partition. Solutions possessing the same surface-tension, though not in molecular equilibrium, would thus be always in osmotic equilibrium. We must not conceal from ourselves that this result would be in contradiction with the kinetic theory.

§ 3. APPLICATION TO THE THEORY OF SOLUTION

If there really exist partitions permeable to one body and impermeable to another, it may be imagined that the homogeneous mixture of these two bodies might be effected in the converse way. It can be easily conceived, in fact, that by the aid of osmotic pressure it would be possible, for example, to dilute or concentrate a solution by driving through the partition in one direction or another a certain quantity of the solvent by means of a pressure kept equal to the osmotic pressure. This is the important fact which Professor Van t' Hoff perceived. The existence of such a wall in all possible cases evidently remains only a very legitimate hypothesis,—a fact which ought not to be concealed.

Relying solely on this postulate, Professor Van t' Hoff easily established, by the most correct method, certain properties of the solutions of gases in a volatile liquid, or of non-volatile bodies in a volatile liquid. To state precisely the other relations, we must admit, in addition, the experimental laws discovered by Pfeffer. But without any hypothesis it becomes possible to demonstrate the laws of Raoult on the lowering of the vapour-tension and of the freezing point of solutions, and also the ratio which connects the heat of fusion with this decrease.

These considerable results can evidently be invoked as *a posteriori* proofs of the exactitude of the experimental laws of osmosis. They are not, however, the only ones that Professor Van t' Hoff has obtained by the same method. This illustrious scholar was thus able to find anew Guldberg and Waage's law on chemical equilibrium at a constant temperature, and to show how the position of the equilibrium changes when the temperature happens to change.

If now we state, in conformity with the laws of Pfeffer, that the product of the osmotic pressure by the volume of the solution is equal to the absolute temperature multiplied by a coefficient, and then look for the numerical figure of this latter in a solution of sugar, for instance, we find that this value is the same as that of the analogous coefficient of the characteristic equation of a perfect gas. There is in this a coincidence which has also been utilized in the preceding thermodynamic calculations. It may be purely fortuitous, but we can hardly refrain from finding in it a physical meaning.

Professor Van t'Hoff has considered this coincidence a demonstration that there exists a strong analogy between a body in solution and a gas; as a matter of fact, it may seem that, in a solution, the distance between the molecules becomes comparable to the molecular distances met with in gases, and that the molecule acquires the same degree of liberty and the same simplicity in both phenomena. In that case it seems probable that solutions will be subject to laws independent of the chemical nature of the dissolved molecule and comparable to the laws governing gases, while if we adopt the kinetic image for the gas, we shall be led to represent to ourselves in a similar way the phenomena which manifest themselves in a solution. Osmotic pressure will then appear to be due to the shock of the dissolved molecules against the membrane. It will come from one side of this partition to superpose itself on the hydrostatic pressure, which latter must have the same value on both sides.

The analogy with a perfect gas naturally becomes much greater as the solution becomes more diluted. It then imitates gas in some other properties; the internal work of the variation of volume is nil, and the specific heat is only a function of the temperature. A solution which is diluted by a reversible method is cooled like a gas which expands adiabatically. [17]

It must, however, be acknowledged that, in other points, the analogy is much less perfect. The opinion which sees in solution a phenomenon resembling fusion, and which has left an indelible trace in everyday language (we shall always say: to melt sugar in water) is certainly not without foundation. Certain of the reasons which might be invoked to uphold this opinion are too evident to be repeated here, though others more recondite might be quoted. The fact that the internal energy generally becomes independent of the concentration when the dilution reaches even a moderately high value is rather in favour of the hypothesis of fusion.

We must not forget, however, the continuity of the liquid and gaseous states; and we may consider it in an absolute way a question devoid of sense to ask whether in a solution the solute is in the liquid or the gaseous state. It is in the fluid state, and perhaps in conditions opposed to those of a body in the state of a perfect gas. It is known, of course, that in this case the manometrical pressure must be regarded as very great in relation to the internal pressure which, in the characteristic equation, is added to the other. May it not seem possible that in the solution it is, on the contrary, the internal pressure which is dominant, the manometric pressure becoming of no account? The coincidence of the formulas would thus be verified, for all the characteristic equations are symmetrical with regard to these two pressures. From this point of view the osmotic pressure would be considered as the result of an attraction between the solvent and the solute; and it would represent the difference between the internal pressures of the solution and of the pure solvent. These hypotheses are highly interesting, and very suggestive; but from the way in which the facts have been set forth, it will appear, no doubt, that there is no obligation to admit them in order to believe in the legitimacy of the application of thermodynamics to the phenomena of solution.

§ 4. ELECTROLYTIC DISSOCIATION

From the outset Professor Van t' Hoff was brought to acknowledge that a great number of solutions formed very notable exceptions which were very irregular in appearance. The analogy with gases did not seem to be maintained, for the osmotic pressure had a very different value from that indicated by the theory. Everything, however, came right if one multiplied by a factor, determined according to each case, but greater than unity, the constant of the characteristic formula. Similar divergences were manifested in the delays observed in congelation, and disappeared when subjected to an analogous correction.

Thus the freezing-point of a normal solution, containing a molecule gramme (that is, the number of grammes equal to the figure representing the molecular mass) of alcohol or sugar in water, falls 1.85° C. If the laws of solution were identically the same for a solution of sea-salt, the same depression should be noticed in a saline solution also containing 1 molecule per litre. In fact, the fall reaches 3.26°, and the solution behaves as if it contained, not 1, but 1.75 normal molecules per litre. The consideration of the osmotic pressures would lead to similar observations, but we know that the experiment would be more difficult and less precise.

We may wonder whether anything really analogous to this can be met with in the case of a gas, and we are thus led to consider the phenomena of dissociation. [18] If we heat a body which, in a gaseous state, is capable of dissociation—hydriodic acid, for example—at a given temperature, an equilibrium is established between three gaseous bodies, the acid, the iodine, and the hydrogen. The total mass will follow with fair closeness Mariotte's law, but the characteristic constant will no longer be the same as in the case of a non-dissociated gas. We here no longer have to do with a single molecule, since each molecule is in part dissociated.

The comparison of the two cases leads to the employment of a new image for representing the phenomenon which has been produced throughout the saline solution. We have introduced a single molecule of salt, and everything occurs as if there were 1.75 molecules. May it not really be said that the number is 1.75, because the sea-salt is partly dissociated, and a molecule has become transformed into 0.75 molecule of sodium, 0.75 of chlorium, and 0.25 of salt?

This is a way of speaking which seems, at first sight, strangely contradicted by experiment. Professor Van t' Hoff, like other chemists, would certainly have rejected—in fact, he did so at first—such a conception, if, about the same time, an illustrious Swedish scholar, M. Arrhenius, had not been brought to the same idea by another road, and, had not by stating it precisely and modifying it, presented it in an acceptable form.

A brief examination will easily show that all the substances which are exceptions to the laws of Van t'Hoff are precisely those which are capable of conducting electricity when undergoing decomposition—that is to say, are electrolytes. The coincidence is absolute, and cannot be simply due to chance.

Now, the phenomena of electrolysis have, for a long time, forced upon us an almost necessary image. The saline molecule is always decomposed, as we know, in the primary phenomenon of electrolysis into two elements which Faraday termed ions. Secondary reactions, no doubt, often come to complicate the question, but these are chemical reactions belonging to the general order of things, and have nothing to do with the electric action working on the solution. The simple phenomenon is always the same—decomposition into two ions, followed by the appearance of one of these ions at the positive and of the other at the negative electrode. But as the very slightest expenditure of energy is sufficient to produce the commencement of electrolysis, it is necessary to suppose that these two ions are not united by any force. Thus the two ions are, in a way, dissociated. Clausius, who was the first to represent the phenomena by this symbol, supposed, in order not to shock the feelings of chemists too much, that this dissociation only affected an infinitesimal fraction of the total number of the molecules of the salt, and thereby escaped all check.

This concession was unfortunate, and the hypothesis thus lost the greater part of its usefulness. M. Arrhenius was bolder, and frankly recognized that dissociation occurs at once in the case of a great number of molecules, and tends to increase more and more as the solution becomes more dilute. It follows the comparison with a gas which, while partially dissociated in an enclosed space, becomes wholly so in an infinite one.

M. Arrhenius was led to adopt this hypothesis by the examination of experimental results relating to the conductivity of electrolytes. In order to interpret certain facts, it has to be recognized that a part only of the molecules in a saline solution can be considered as conductors of electricity, and that by adding water the number of molecular conductors is increased. This increase, too, though rapid at first, soon becomes slower, and approaches a certain limit which an infinite dilution would enable it to attain. If the conducting molecules are the dissociated molecules, then the dissociation (so long as it is a question of strong acids and salts) tends to become complete in the case of an unlimited dilution.

The opposition of a large number of chemists and physicists to the ideas of M. Arrhenius was at first very fierce. It must be noted with regret that, in France particularly, recourse was had to an arm which scholars often wield rather clumsily. They joked about these free ions in solution, and they asked to see this chlorine and this sodium which swam about the water in a state of liberty. But in science, as elsewhere, irony is not argument, and it soon had to be acknowledged that the hypothesis of M. Arrhenius showed itself singularly fertile and had to be regarded, at all events, as a very expressive image, if not, indeed, entirely in conformity with reality.

It would certainly be contrary to all experience, and even to common sense itself, to suppose that in dissolved chloride of sodium there is really free sodium, if we suppose these atoms of sodium to be absolutely identical with ordinary atoms. But there is a great difference. In the one case the atoms are electrified, and carry a relatively considerable positive charge, inseparable from their state as ions, while in the other they are in the neutral state. We may suppose that the presence of this charge brings about modifications as extensive as one pleases in the chemical properties of the atom. Thus the hypothesis will be removed from all discussion of a chemical order, since it will have been made plastic enough beforehand to adapt itself to all the known facts; and if we object that sodium cannot subsist in water because it instantaneously decomposes the latter, the answer is simply that the sodium ion does not decompose water as does ordinary sodium.

Still, other objections might be raised which could not be so easily refuted. One, to which chemists not unreasonably attached great importance, was this:—If a certain quantity of chloride of sodium is dissociated into chlorine and sodium, it should be possible, by diffusion, for example, which brings out plainly the phenomena of dissociation in gases, to extract from the solution a part either of the chlorine or of the sodium, while the corresponding part of the other compound would remain. This result would be in flagrant contradiction with the fact that, everywhere and always, a solution of salt contains strictly the same proportions of its component elements.

M. Arrhenius answers to this that the electrical forces in ordinary conditions prevent separation by diffusion or by any other process. Professor Nernst goes further, and has shown that the concentration currents which are produced when two electrodes of the same substance are plunged into two unequally concentrated solutions may be interpreted by the hypothesis that, in these particular conditions, the diffusion does bring about a separation of the ions. Thus the argument is turned round, and the proof supposed to be given of the incorrectness of the theory becomes a further reason in its favour.

It is possible, no doubt, to adduce a few other experiments which are not very favourable to M. Arrhenius's point of view, but they are isolated cases; and, on the whole, his theory has enabled many isolated facts, till then scattered, to be co-ordinated, and has allowed very varied phenomena to be linked together. It has also suggested—and, moreover, still daily suggests—researches of the highest order.

In the first place, the theory of Arrhenius explains electrolysis very simply. The ions which, so to speak, wander about haphazard, and are uniformly distributed throughout the liquid, steer a regular course as soon as we dip in the trough containing the electrolyte the two electrodes connected with the poles of the dynamo or generator of electricity. Then the charged positive ions travel in the direction of the electromotive force and the negative ions in the opposite direction. On reaching the electrodes they yield up to them the charges they carry, and thus pass from the state of ion into that of ordinary atom. Moreover, for the solution to remain in equilibrium, the vanished ions must be immediately replaced by others, and thus the state of ionisation of the electrolyte remains constant and its conductivity persists.

All the peculiarities of electrolysis are capable of interpretation: the phenomena of the transport of ions, the fine experiments of M. Bouty, those of Professor Kohlrausch and of Professor Ostwald on various points in electrolytic conduction, all support the theory. The verifications of it can even be quantitative, and we can foresee numerical relations between conductivity and other phenomena. The measurement of the conductivity permits the number of molecules dissociated in a given solution to be calculated, and the number is thus found to be precisely the same as that arrived at if it is wished to remove the disagreement between reality and the anticipations which result from the theory of Professor Van t' Hoff. The laws of cryoscopy, of tonometry, and of osmosis thus again become strict, and no exception to them remains.

If the dissociation of salts is a reality and is complete in a dilute solution, any of the properties of a saline solution whatever should be represented numerically as the sum of three values, of which one concerns the positive ion, a second the negative ion, and the third the solvent. The properties of the solutions would then be what are called additive properties. Numerous verifications may be attempted by very different roads. They generally succeed very well; and whether we measure the electric conductivity, the density, the specific heats, the index of refraction, the power of rotatory polarization, the colour, or the absorption spectrum, the additive property will everywhere be found in the solution.

The hypothesis, so contested at the outset by the chemists, is, moreover, assuring its triumph by important conquests in the domain of chemistry itself. It permits us to give a vivid explanation of chemical reaction, and for the old motto of the chemists, "Corpora non agunt, nisi soluta," it substitutes a modern one, "It is especially the ions which react." Thus, for example, all salts of iron, which contain iron in the state of ions, give similar reactions; but salts such as ferrocyanide of potassium, in which iron does not play the part of an ion, never give the characteristic reactions of iron.

Professor Ostwald and his pupils have drawn from the hypothesis of Arrhenius manifold consequences which have been the cause of considerable progress in physical chemistry. Professor Ostwald has shown, in particular, how this hypothesis permits the quantitative calculation of the conditions of equilibrium of electrolytes and solutions, and especially of the phenomena of neutralization. If a dissolved salt is partly dissociated into ions, this solution must be limited by an equilibrium between the non-dissociated molecule and the two ions resulting from the dissociation; and, assimilating the phenomenon to the case of gases, we may take for its study the laws of Gibbs and of Guldberg and Waage. The results are generally very satisfactory, and new researches daily furnish new checks.

Professor Nernst, who before gave, as has been said, a remarkable interpretation of the diffusion of electrolytes, has, in the direction pointed out by M. Arrhenius, developed a theory of the entire phenomena of electrolysis, which, in particular, furnishes a striking explanation of the mechanism of the production of electromotive force in galvanic batteries.

Extending the analogy, already so happily invoked, between the phenomena met with in solutions and those produced in gases, Professor Nernst supposes that metals tend, as it were, to vaporize when in presence of a liquid. A piece of zinc introduced, for example, into pure water gives birth to a few metallic ions. These ions become positively charged, while the metal naturally takes an equal charge, but of contrary sign. Thus the solution and the metal are both electrified; but this sort of vaporization is hindered by electrostatic attraction, and as the charges borne by the ions are considerable, an equilibrium will be established, although the number of ions which enter the solution will be very small.

If the liquid, instead of being a solvent like pure water, contains an electrolyte, it already contains metallic ions, the osmotic pressure of which will be opposite to that of the solution. Three cases may then present themselves—either there will be equilibrium, or the electrostatic attraction will oppose itself to the pressure of solution and the metal will be negatively charged, or, finally, the attraction will act in the same direction as the pressure, and the metal will become positively and the solution negatively charged. Developing this idea, Professor Nernst calculates, by means of the action of the osmotic pressures, the variations of energy brought into play and the value of the differences of potential by the contact of the electrodes and electrolytes. He deduces this from the electromotive force of a single battery cell which becomes thus connected with the values of the osmotic pressures, or, if you will, thanks to the relation discovered by Van t' Hoff, with the concentrations. Some particularly interesting electrical phenomena thus become connected with an already very important group, and a new bridge is built which unites two regions long considered foreign to each other.

The recent discoveries on the phenomena produced in gases when rendered conductors of electricity almost force upon us, as we shall see, the idea that there exist in these gases electrified centres moving through the field, and this idea gives still greater probability to the analogous theory explaining the mechanism of the conductivity of liquids. It will also be useful, in order to avoid confusion, to restate with precision this notion of electrolytic ions, and to ascertain their magnitude, charge, and velocity.

The two classic laws of Faraday will supply us with important information. The first indicates that the quantity of electricity passing through the liquid is proportional to the quantity of matter deposited on the electrodes. This leads us at once to the consideration that, in any given solution, all the ions possess individual charges equal in absolute value.

The second law may be stated in these terms: an atom-gramme of metal carries with it into electrolysis a quantity of electricity proportionate to its valency. [19]

Numerous experiments have made known the total mass of hydrogen capable of carrying one coulomb, and it will therefore be possible to estimate the charge of an ion of hydrogen if the number of atoms of hydrogen in a given mass be known. This last figure is already furnished by considerations derived from the kinetic theory, and agrees with the one which can be deduced from the study of various phenomena. The result is that an ion of hydrogen having a mass of 1.3×10^{-20} grammes bears a charge of 1.3×10^{-20} electromagnetic units; and the second law will immediately enable the charge of any other ion to be similarly estimated.

The measurements of conductivity, joined to certain considerations relating to the differences of concentration which appear round the electrode in electrolysis, allow the speed of the ions to be calculated. Thus, in a liquid containing 1/10th of a hydrogen-ion per litre, the absolute speed of an ion would be 3/10ths of a millimetre per second in a field where the fall of potential would be 1 volt per centimetre. Sir Oliver Lodge, who has made direct experiments to measure this speed, has obtained a figure very approximate to this. This value is very small compared to that which we shall meet with in gases.

Another consequence of the laws of Faraday, to which, as early as 1881, Helmholtz drew attention, may be considered as the starting-point of certain new doctrines we shall come across later.

Helmholtz says: "If we accept the hypothesis that simple bodies are composed of atoms, we are obliged to admit that, in the same way, electricity, whether positive or negative, is composed of elementary parts which behave like atoms of electricity."

The second law seems, in fact, analogous to the law of multiple proportions in chemistry, and it shows us that the quantities of electricity carried vary from the simple to the double or treble, according as it is a question of a uni-, bi-, or trivalent metal; and as the chemical law leads up to the conception of the material atom, so does the electrolytic law suggest the idea of an electric atom.

CHAPTER VI
THE ETHER

§ 1. THE LUMINIFEROUS ETHER

It is in the works of Descartes that we find the first idea of attributing those physical phenomena which the properties of matter fail to explain to some subtle matter which is the receptacle of the energy of the universe.

In our times this idea has had extraordinary luck. After having been eclipsed for two hundred years by the success of the immortal synthesis of Newton, it gained an entirely new splendour with Fresnel and his followers. Thanks to their admirable discoveries, the first stage seemed accomplished, the laws of optics were represented by a single hypothesis, marvellously fitted to allow us to anticipate unknown phenomena, and all these anticipations were subsequently fully verified by experiment. But the researches of Faraday, Maxwell, and Hertz authorized still greater ambitions; and it really seemed that this medium, to which it was agreed to give the ancient name of ether, and which had already explained light and radiant heat, would also be sufficient to explain electricity. Thus the hope began to take form that we might succeed in demonstrating the unity of all physical forces. It was thought that the knowledge of the laws relating to the inmost movements of this ether might give us the key to all phenomena, and might make us acquainted with the method in which energy is stored up, transmitted, and parcelled out in its external manifestations.

We cannot study here all the problems which are connected with the physics of the ether. To do this a complete treatise on optics would have to be written and a very lengthy one on electricity. I shall simply endeavour to show rapidly how in the last few years the ideas relative to the constitution of this ether have evolved, and we shall see if it be possible without self-delusion to imagine that a single medium can really allow us to group all the known facts in one comprehensive arrangement. As constructed by Fresnel, the hypothesis of the luminous ether, which had so great a struggle at the outset to overcome the stubborn resistance of the partisans of the then classic theory of emission, seemed, on the contrary, to possess in the sequel an unshakable strength. Lamé, though a prudent mathematician, wrote: "*The existence* of the ethereal fluid is *incontestably demonstrated* by the propagation of light through the planetary spaces, and by the explanation, so simple and so complete, of the phenomena of diffraction in the wave theory of light"; and he adds: "The laws of double refraction prove with no less certainty that the *ether exists* in all diaphanous media." Thus the ether was no longer an hypothesis, but in some sort a tangible reality. But the ethereal fluid of which the existence was thus proclaimed has some singular properties.

Were it only a question of explaining rectilinear propagation, reflexion, refraction, diffraction, and interferences notwithstanding grave difficulties at the outset and the objections formulated by Laplace and Poisson (some of which, though treated somewhat lightly at the present day, have not lost all value), we should be under no obligation to make any hypothesis other than that of the undulations of an elastic medium, without deciding in advance anything as to the nature and direction of the vibrations.

This medium would, naturally—since it exists in what we call the void—be considered as imponderable. It may be compared to a fluid of negligible mass—since it offers no appreciable resistance to the motion of the planets—but is endowed with an enormous elasticity, because the velocity of the propagation of light is considerable. It must be capable of penetrating into all transparent bodies, and of retaining there, so to speak, a constant elasticity, but must there become condensed, since the speed of propagation in these bodies is less than in a vacuum. Such properties belong to no material gas, even the most rarefied, but they admit of no essential contradiction, and that is the important point. [20]

It was the study of the phenomena of polarization which led Fresnel to his bold conception of transverse vibrations, and subsequently induced him to penetrate further into the constitution of the ether. We know the experiment of Arago on the noninterference of polarized rays in rectangular planes. While two systems of waves, proceeding from the same source of natural light and propagating themselves in nearly parallel directions, increase or become destroyed according to whether the nature of the superposed waves are of the same or of contrary signs, the waves of the rays polarized in perpendicular planes, on the other hand, can never interfere with each other. Whatever the difference of their course, the intensity of the light is always the sum of the intensity of the two rays.

Fresnel perceived that this experiment absolutely compels us to reject the hypothesis of longitudinal vibrations acting along the line of propagation in the direction of the rays. To explain it, it must of necessity be admitted, on the contrary, that the vibrations are transverse and perpendicular to the ray. Verdet could say, in all truth, "It is not possible to deny the transverse direction of luminous vibrations, without at the same time denying that light consists of an undulatory movement."

Such vibrations do not and cannot exist in any medium resembling a fluid. The characteristic of a fluid is that its different parts can displace themselves with regard to one another without any reaction appearing so long as a variation of volume is not produced. There certainly may exist, as we have seen, certain traces of rigidity in a liquid, but we cannot conceive such a thing in a body infinitely more subtle than rarefied gas. Among material bodies, a solid alone really possesses the rigidity sufficient for the production within it of transverse vibrations and for their maintenance during their propagation.

Since we have to attribute such a property to the ether, we may add that on this point it resembles a solid, and Lord Kelvin has shown that this solid, would be much more rigid than steel. This conclusion produces great surprise in all who hear it for the first time, and it is not rare to hear it appealed to as an argument against the actual existence of the ether. It does not seem, however, that such an argument can be decisive. There is no reason for supposing that the ether ought to be a sort of extension of the bodies we are accustomed to handle. Its properties may astonish our ordinary way of thinking, but this rather unscientific astonishment is not a reason for doubting its existence. Real difficulties would appear only if we were led to attribute to the ether, not singular properties which are seldom found united in the same substance, but properties logically contradictory. In short, however odd such a medium may appear to us, it cannot be said that there is any absolute incompatibility between its attributes.

It would even be possible, if we wished, to suggest images capable of representing these contrary appearances. Various authors have done so. Thus, M. Boussinesq assumes that the ether behaves like a very rarefied gas in respect of the celestial bodies, because these last move, while bathed in it, in all directions and relatively slowly, while they permit it to retain, so to speak, its perfect homogeneity. On the other hand, its own undulations are so rapid that so far as they are concerned the conditions become very different, and its fluidity has, one might say, no longer the time to come in. Hence its rigidity alone appears.

Another consequence, very important in principle, of the fact that vibrations of light are transverse, has been well put in evidence by Fresnel. He showed how we have, in order to understand the action which excites without condensation the sliding of successive layers of the ether during the propagation of a vibration, to consider the vibrating medium as being composed of molecules separated by finite distances. Certain authors, it is true, have proposed theories in which the action at a distance of these molecules are replaced by actions of contact between parallelepipeds sliding over one another; but, at bottom, these two points of view both lead us to conceive the ether as a discontinuous medium, like matter itself. The ideas gathered from the most recent experiments also bring us to the same conclusion.

§ 2. RADIATIONS

In the ether thus constituted there are therefore propagated transverse vibrations, regarding which all experiments in optics furnish very precise information. The amplitude of these vibrations is exceedingly small, even in relation to the wave-length, small as these last are. If, in fact, the amplitude of the vibrations acquired a noticeable value in comparison with the wave-length, the speed of propagation should increase with the amplitude. Yet, in spite of some curious experiments which seem to establish that the speed of light does alter a little with its intensity, we have reason to believe that, as regards light, the amplitude of the oscillations in relation to the wave-length is incomparably less than in the case of sound.

It has become the custom to characterise each vibration by the path which the vibratory movement traverses during the space of a vibration—by the length of wave, in a word—rather than by the duration of the vibration itself. To measure wave-lengths, the methods must be employed to which I have already alluded on the subject of measurements of length. Professor Michelson, on the one hand, and MM. Perot and Fabry, on the other, have devised exceedingly ingenious processes, which have led to results of really unhoped-for precision. The very exact knowledge also of the speed of the propagation of light allows the duration of a vibration to be calculated when once the wave-length is known. It is thus found that, in the case of visible light, the number of the vibrations from the end of the violet to the infra-red varies from four hundred to two hundred billions per second. This gamut is not, however, the only one the ether can give. For a long time we have known ultra-violet radiations still more rapid, and, on the other hand, infra-red ones more slow, while in the last few years the field of known radiations has been singularly extended in both directions.

It is to M. Rubens and his fellow-workers that are due the most brilliant conquests in the matter of great wave-lengths. He had remarked that, in their study, the difficulty of research proceeds from the fact that the extreme waves of the infra-red spectrum only contain a small part of the total energy emitted by an incandescent body; so that if, for the purpose of study, they are further dispersed by a prism or a grating, the intensity at any one point becomes so slight as to be no longer observable. His original idea was to obtain, without prism or grating, a homogeneous pencil of great wave-length sufficiently intense to be examined. For this purpose the radiant source used was a strip of platinum covered with fluorine or powdered quartz, which emits numerous radiations close to two bands of linear absorption in the absorption spectra of fluorine and quartz, one of which is situated in the infra-red. The radiations thus emitted are several times reflected on fluorine or on quartz, as the case may be; and as, in proximity to the bands, the absorption is of the order of that of metallic bodies for luminous rays, we no longer meet in the pencil several times reflected or in the rays *remaining* after this kind of filtration, with any but radiations of great wave-length. Thus, for instance, in the case of the quartz, in the neighbourhood of a radiation corresponding to a wave-length of 8.5 microns, the absorption is thirty times greater in the region of the band than in the neighbouring region, and consequently, after three reflexions, while the corresponding radiations will not have been weakened, the neighbouring waves will be so, on the contrary, in the proportion of 1 to 27,000.

With mirrors of rock salt and of sylvine[21] there have been obtained, by taking an incandescent gas light (Auer) as source, radiations extending as far as 70 microns; and these last are the greatest wave-lengths observed in optical phenomena. These radiations are largely absorbed by the vapour of water, and it is no doubt owing to this absorption that they are not found in the solar spectrum. On the other hand, they easily pass through gutta-percha, india-rubber, and insulating substances in general.

At the opposite end of the spectrum the knowledge of the ultra-violet regions has been greatly extended by the researches of Lenard. These extremely rapid radiations have been shown by that eminent physicist to occur in the light of the electric sparks which flash between two metal points, and which are produced by a large induction coil with condenser and a Wehnelt break. Professor Schumann has succeeded in photographing them by depositing bromide of silver directly on glass plates without fixing it with gelatine; and he has, by the same process, photographed in the spectrum of hydrogen a ray with a wave-length of only 0.1 micron.

The spectroscope was formed entirely of fluor-spar, and a vacuum had been created in it, for these radiations are extremely absorbable by the air.

Notwithstanding the extreme smallness of the luminous wave-lengths, it has been possible, after numerous fruitless trials, to obtain stationary waves analogous to those which, in the case of sound, are produced in organ pipes. The marvellous application M. Lippmann has made of these waves to completely solve the problem of photography in colours is well known. This discovery, so important in itself and so instructive, since it shows us how the most delicate anticipations of theory may be verified in all their consequences, and lead the physicist to the solution of the problems occurring in practice, has justly become popular, and there is, therefore, no need to describe it here in detail.

Professor Wiener obtained stationary waves some little while before M. Lippmann's discovery, in a layer of a sensitive substance having a grain sufficiently small in relation to the length of wave. His aim was to solve a question of great importance to a complete knowledge of the ether. Fresnel founded his theory of double refraction and reflexion by transparent surfaces, on the hypothesis that the vibration of a ray of polarized light is perpendicular to the plane of polarization. But Neumann has proposed, on the contrary, a theory in which he recognizes that the luminous vibration is in this very plane. He rather supposes, in opposition to Fresnel's idea, that the density of the ether remains the same in all media, while its coefficient of elasticity is variable.

Very remarkable experiments on dispersion by M. Carvallo prove indeed that the idea of Fresnel was, if not necessary for us to adopt, at least the more probable of the two; but apart from this indication, and contrary to the hypothesis of Neumann, the two theories, from the point of view of the explanation of all known facts, really appear to be equivalent. Are we then in presence of two mechanical explanations, different indeed, but nevertheless both adaptable to all the facts, and between which it will always be impossible to make a choice? Or, on the contrary, shall we succeed in realising an *experimentum crucis*, an experiment at the point where the two theories cross, which will definitely settle the question?

Professor Wiener thought he could draw from his experiment a firm conclusion on the point in dispute. He produced stationary waves with light polarized at an angle of 45°,[22] and established that, when light is polarized in the plane of incidence, the fringes persist; but that, on the other hand, they disappear when the light is polarized perpendicularly to this plane. If it be admitted that a photographic impression results from the active force of the vibratory movement of the ether, the question is, in fact, completely elucidated, and the discrepancy is abolished in Fresnel's favour.

M.H. Poincaré has pointed out, however, that we know nothing as to the mechanism of the photographic impression. We cannot consider it evident that it is the kinetic energy of the ether which produces the decomposition of the sensitive salt; and if, on the contrary, we suppose it to be due to the potential energy, all the conclusions are reversed, and Neumann's idea triumphs.

Recently a very clever physicist, M. Cotton, especially known for his skilful researches in the domain of optics, has taken up anew the study of stationary waves. He has made very precise quantitative experiments, and has demonstrated, in his turn, that it is impossible, even with spherical waves, to succeed in determining on which of the two vectors which have to be regarded in all theories of light on the subject of polarization phenomena the luminous intensity and the chemical action really depend. This question, therefore, no longer exists for those physicists who admit that luminous vibrations are electrical oscillations. Whatever, then, the hypothesis formed, whether it be electric force or, on the contrary, magnetic force which we place in the plane of polarization, the mode of propagation foreseen will always be in accord with the facts observed.

§ 3. THE ELECTROMAGNETIC ETHER

The idea of attributing the phenomena of electricity to perturbations produced in the medium which transmits the light is already of old standing; and the physicists who witnessed the triumph of Fresnel's theories could not fail to conceive that this fluid, which fills the whole of space and penetrates into all bodies, might also play a preponderant part in electrical actions. Some even formed too hasty hypotheses on this point; for the hour had not arrived when it was possible to place them on a sufficiently sound basis, and the known facts were not numerous enough to give the necessary precision.

The founders of modern electricity also thought it wiser to adopt, with regard to this science, the attitude taken by Newton in connection with gravitation: "In the first place to observe facts, to vary the circumstances of these as much as possible, to accompany this first work by precise measurements in order to deduce from them general laws founded solely on experiment, and to deduce from these laws, independently of all hypotheses on the nature of the forces producing the phenomena, the mathematical value of these forces—that is to say, the formula representing them. Such was the system pursued by Newton. It has, in general, been adopted in France by the scholars to whom physics owe the great progress made of late years, and it has served as my guide in all my researches on electrodynamic phenomena.... It is for this reason that I have avoided speaking of the ideas I may have on the nature of the cause of the force emanating from voltaic conductors."

Thus did Ampère express himself. The illustrious physicist rightly considered the results obtained by him through following this wise method as worthy of comparison with the laws of attraction; but he knew that when this first halting-place was reached there was still further to go, and that the evolution of ideas must necessarily continue.

"With whatever physical cause," he adds, "we may wish to connect the phenomena produced by electro-dynamic action, the formula I have obtained will always remain the expression of the facts," and he explicitly indicated that if one could succeed in deducing his formula from the consideration of the vibrations of a fluid distributed through space, an enormous step would have been taken in this department of physics. He added, however, that this research appeared to him premature, and would change nothing in the results of his work, since, to accord with facts, the hypothesis adopted would always have to agree with the formula which exactly represents them.

It is not devoid of interest to observe that Ampère himself, notwithstanding his caution, really formed some hypotheses, and recognized that electrical phenomena were governed by the laws of mechanics. Yet the principles of Newton then appeared to be unshakable.

Faraday was the first to demonstrate, by clear experiment, the influence of the media in electricity and magnetic phenomena, and he attributed this influence to certain modifications in the ether which these media enclose. His fundamental conception was to reject action at a distance, and to localize in the ether the energy whose evolution is the cause of the actions manifested, as, for example, in the discharge of a condenser.

Consider the barrel of a pump placed in a vacuum and closed by a piston at each end, and let us introduce between these a certain mass of air. The two pistons, through the elastic force of the gas, repel each other with a force which, according to the law of Mariotte, varies in inverse ratio to the distance. The method favoured by Ampère would first of all allow this law of repulsion between the two pistons to be discovered, even if the existence of a gas enclosed in the barrel of the pump were unsuspected; and it would then be natural to localize the potential energy of the system on the surface of the two pistons. But if the phenomenon is more carefully examined, we shall discover the presence of the air, and we shall understand that every part of the volume of this air could, if it were drawn off into a recipient of equal volume, carry away with it a fraction of the energy of the system, and that consequently this energy belongs really to the air and not to the pistons, which are there solely for the purpose of enabling this energy to manifest its existence.

Faraday made, in some sort, an equivalent discovery when he perceived that the electrical energy belongs, not to the coatings of the condenser, but to the dielectric which separates them. His audacious views revealed to him a new world, but to explore this world a surer and more patient method was needed.

Maxwell succeeded in stating with precision certain points of Faraday's ideas, and he gave them the mathematical form which, often wrongly, impresses physicists, but which when it exactly encloses a theory, is a certain proof that this theory is at least coherent and logical. [23]

The work of Maxwell is over-elaborated, complex, difficult to read, and often ill-understood, even at the present day. Maxwell is more concerned in discovering whether it is possible to give an explanation of electrical and magnetic phenomena which shall be founded on the mechanical properties of a single medium, than in stating this explanation in precise terms. He is aware that if we could succeed in constructing such an interpretation, it would be easy to propose an infinity of others, entirely equivalent from the point of view of the experimentally verifiable consequences; and his especial ambition is therefore to extract from the premises a general view, and to place in evidence something which would remain the common property of all the theories.

He succeeded in showing that if the electrostatic energy of an electromagnetic field be considered to represent potential energy, and its electrodynamic the kinetic energy, it becomes possible to satisfy both the principle of least action and that of the conservation of energy; from that moment—if we eliminate a few difficulties which exist regarding the stability of the solutions—the possibility of finding mechanical explanations of electromagnetic phenomena must be considered as demonstrated. He thus succeeded, moreover, in stating precisely the notion of two electric and magnetic fields which are produced in all points of space, and which are strictly inter-connected, since the variation of the one immediately and compulsorily gives birth to the other.

From this hypothesis he deduced that, in the medium where this energy is localized, an electromagnetic wave is propagated with a velocity equal to the relation of the units of electric mass in the electromagnetic and electrostatic systems. Now, experiments made known since his time have proved that this relation is numerically equal to the speed of light, and the more precise experiments made in consequence—among which should be cited the particularly careful ones of M. Max Abraham—have only rendered the coincidence still more complete.

It is natural henceforth to suppose that this medium is identical with the luminous ether, and that a luminous wave is an electromagnetic wave—that is to say, a succession of alternating currents, which exist in the dielectric and even in the void, and possess an enormous frequency, inasmuch as they change their direction thousands of billions of times per second, and by reason of this frequency produce considerable induction effects. Maxwell did not admit the existence of open currents. To his mind, therefore, an electrical vibration could not produce condensations of electricity. It was, in consequence, necessarily transverse, and thus coincided with the vibration of Fresnel; while the corresponding magnetic vibration was perpendicular to it, and would coincide with the luminous vibration of Neumann.

Maxwell's theory thus establishes a close correlation between the phenomena of the luminous and those of the electromagnetic waves, or, we might even say, the complete identity of the two. But it does not follow from this that we ought to regard the variation of an electric field produced at some one point as necessarily consisting of a real displacement of the ether round that point. The idea of thus bringing electrical phenomena back to the mechanics of the ether is not, then, forced upon us, and the contrary idea even seems more probable. It is not the optics of Fresnel which absorbs the science of electricity, it is rather the optics which is swallowed up by a more general theory. The attempts of popularizers who endeavour to represent, in all their details, the mechanism of the electric phenomena, thus appear vain enough, and even puerile. It is useless to find out to what material body the ether may be compared, if we content ourselves with seeing in it a medium of which, at every point, two vectors define the properties.

For a long time, therefore, we could remark that the theory of Fresnel simply supposed a medium in which something periodical was propagated, without its being necessary to admit this something to be a movement; but we had to wait not only for Maxwell, but also for Hertz, before this idea assumed a really scientific shape. Hertz insisted on the fact that the six equations of the electric field permit all the phenomena to be anticipated without its being necessary to construct one hypothesis or another, and he put these equations into a very symmetrical form, which brings completely in evidence the perfect reciprocity between electrical and magnetic actions. He did yet more, for he brought to the ideas of Maxwell the most striking confirmation by his memorable researches on electric oscillations.

§ 4. ELECTRICAL OSCILLATIONS

The experiments of Hertz are well known. We know how the Bonn physicist developed, by means of oscillating electric discharges, displacement currents and induction effects in the whole of the space round the spark-gap; and how he excited by induction at some point in a wire a perturbation which afterwards is propagated along the wire, and how a resonator enabled him to detect the effect produced.

The most important point made evident by the observation of interference phenomena and subsequently verified directly by M. Blondlot, is that the electromagnetic perturbation is propagated with the speed of light, and this result condemns for ever all the hypotheses which fail to attribute any part to the intervening media in the propagation of an induction phenomenon.

If the inducing action were, in fact, to operate directly between the inducing and the induced circuits, the propagation should be instantaneous; for if an interval were to occur between the moment when the cause acted and the one when the effect was produced, during this interval there would no longer be anything anywhere, since the intervening medium does not come into play, and the phenomenon would then disappear.

Leaving on one side the manifold but purely electrical consequences of this and the numerous researches relating to the production or to the properties of the waves—some of which, those of MM. Sarrazin and de la Rive, Righi, Turpain, Lebedeff, Decombe, Barbillon, Drude, Gutton, Lamotte, Lecher, etc., are, however, of the highest order—I shall only mention here the studies more particularly directed to the establishment of the identity of the electromagnetic and the luminous waves.

The only differences which subsist are necessarily those due to the considerable discrepancy which exists between the durations of the periods of these two categories of waves. The length of wave corresponding to the first spark-gap of Hertz was about 6 metres, and the longest waves perceptible by the retina are 7/10 of a micron. [24]

These radiations are so far apart that it is not astonishing that their properties have not a perfect similitude. Thus phenomena like those of diffraction, which are negligible in the ordinary conditions under which light is observed, may here assume a preponderating importance. To play the part, for example, with the Hertzian waves, which a mirror 1 millimetre square plays with regard to light, would require a colossal mirror which would attain the size of a myriametre [25] square.

The efforts of physicists have to-day, however, filled up, in great part, this interval, and from both banks at once they have laboured to build a bridge between the two domains. We have seen how Rubens showed us calorific rays 60 metres long; on the other hand, MM. Lecher, Bose, and Lampa have succeeded, one after the other, in gradually obtaining oscillations with shorter and shorter periods. There have been produced, and are now being studied, electromagnetic waves of four millimetres; and the gap subsisting in the spectrum between the rays left undetected by sylvine and the radiations of M. Lampa now hardly comprise more than five octaves—that is to say, an interval perceptibly equal to that which separates the rays observed by M. Rubens from the last which are evident to the eye.

The analogy then becomes quite close, and in the remaining rays the properties, so to speak, characteristic of the Hertzian waves, begin to appear. For these waves, as we have seen, the most transparent bodies are the most perfect electrical insulators; while bodies still slightly conducting are entirely opaque. The index of refraction of these substances tends in the case of great wave-lengths to become, as the theory anticipates, nearly the square root of the dielectric constant.

MM. Rubens and Nichols have even produced with the waves which remain phenomena of electric resonance quite similar to those which an Italian scholar, M. Garbasso, obtained with electric waves. This physicist showed that, if the electric waves are made to impinge on a flat wooden stand, on which are a series of resonators parallel to each other and uniformly arranged, these waves are hardly reflected save in the case where the resonators have the same period as the spark-gap. If the remaining rays are allowed to fall on a glass plate silvered and divided by a diamond fixed on a dividing machine into small rectangles of equal dimensions, there will be observed variations in the reflecting power according to the orientation of the rectangles, under conditions entirely comparable with the experiment of Garbasso.

In order that the phenomenon be produced it is necessary that the remaining waves should be previously polarized. This is because, in fact, the mechanism employed to produce the electric oscillations evidently gives out vibrations which occur on a single plane and are subsequently polarized.

We cannot therefore entirely assimilate a radiation proceeding from a spark-gap to a ray of natural light. For the synthesis of light to be realized, still other conditions must be complied with. During a luminous impression, the direction and the phase change millions of times in the vibration sensible to the retina, yet the damping of this vibration is very slow. With the Hertzian oscillations all these conditions are changed—the damping is very rapid but the direction remains invariable.

Every time, however, that we deal with general phenomena which are independent of these special conditions, the parallelism is perfect; and with the waves, we have put in evidence the reflexion, refraction, total reflexion, double reflexion, rotatory polarization, dispersion, and the ordinary interferences produced by rays travelling in the same direction and crossing each other at a very acute angle, or the interferences analogous to those which Wiener observed with rays of the contrary direction.

A very important consequence of the electromagnetic theory foreseen by Maxwell is that the luminous waves which fall on a surface must exercise on this surface a pressure equal to the radiant energy which exists in the unit of volume of the surrounding space. M. Lebedeff a few years ago allowed a sheaf of rays from an arc lamp to fall on a deflection radiometer, [26] and thus succeeded in revealing the existence of this pressure. Its value is sufficient, in the case of matter of little density and finely divided, to reduce and even change into repulsion the attractive action exercised on bodies by the sun. This is a fact formerly conjectured by Faye, and must certainly play a great part in the deformation of the heads of comets.

More recently, MM. Nichols and Hull have undertaken experiments on this point. They have measured not only the pressure, but also the energy of the radiation by means of a special bolometer. They have thus arrived at numerical verifications which are entirely in conformity with the calculations of Maxwell.

The existence of these pressures may be otherwise foreseen even apart from the electromagnetic theory, by adding to the theory of undulations the principles of thermodynamics. Bartoli, and more recently Dr Larmor, have shown, in fact, that if these pressures did not exist, it would be possible, without any other phenomenon, to pass heat from a cold into a warm body, and thus transgress the principle of Carnot.

§ 5. THE X RAYS

It appears to-day quite probable that the X rays should be classed among the phenomena which have their seat in the luminous ether. Doubtless it is not necessary to recall here how, in December 1895, Röntgen, having wrapped in black paper a Crookes tube in action, observed that a fluorescent platinocyanide of barium screen placed in the neighbourhood, had become visible in the dark, and that a photographic plate had received an impress. The rays which come from the tube, in conditions now well known, are not deviated by a magnet, and, as M. Curie and M. Sagnac have conclusively shown, they carry no electric charge. They are subject to neither reflection nor refraction, and very precise and very ingenious measurements by M. Gouy have shown that, in their case, the refraction index of the various bodies cannot be more than a millionth removed from unity.

We knew from the outset that there existed various X rays differing from each other as, for instance, the colours of the spectrum, and these are distinguished from each other by their unequal power of passing through substances. M. Sagnac, particularly, has shown that there can be obtained a gradually decreasing scale of more or less absorbable rays, so that the greater part of their photographic action is stopped by a simple sheet of black paper. These rays figure among the secondary rays discovered, as is known, by this ingenious physicist. The X rays falling on matter are thus subjected to transformations which may be compared to those which the phenomena of luminescence produce on the ultra-violet rays.

M. Benoist has founded on the transparency of matter to the rays a sure and practical method of allowing them to be distinguished, and has thus been enabled to define a specific character analogous to the colour of the rays of light. It is probable also that the different rays do not transport individually the same quantity of energy. We have not yet obtained on this point precise results, but it is roughly known, since the experiments of MM. Rutherford and M'Clung, what quantity of energy corresponds to a pencil of X rays. These physicists have found that this quantity would be, on an average, five hundred times larger than that brought by an analogous pencil of solar light to the surface of the earth. What is the nature of this energy? The question does not appear to have been yet solved.

It certainly appears, according to Professors Haga and Wind and to Professor Sommerfeld, that with the X rays curious experiments of diffraction may be produced. Dr Barkla has shown also that they can manifest true polarization. The secondary rays emitted by a metallic surface when struck by X rays vary, in fact, in intensity when the position of the plane of incidence round the primary pencil is changed. Various physicists have endeavoured to measure the speed of propagation, but it seems more and more probable that it is very nearly that of light.[27]

I must here leave out the description of a crowd of other experiments. Some very interesting researches by M. Brunhes, M. Broca, M. Colardeau, M. Villard, in France, and by many others abroad, have permitted the elucidation of several interesting problems relative to the duration of the emission or to the best disposition to be adopted for the production of the rays. The only point which will detain us is the important question as to the nature of the X rays themselves; the properties which have just been brought to mind are those which appear essential and which every theory must reckon with.

The most natural hypothesis would be to consider the rays as ultra-violet radiations of very short wave-length, or radiations which are in a manner ultra-ultra-violet. This interpretation can still, at this present moment, be maintained, and the researches of MM. Buisson, Righi, Lenard, and Merrit Stewart have even established that rays of very short wave-lengths produce on metallic conductors, from the point of view of electrical phenomena, effects quite analogous to those of the X rays. Another resemblance results also from the experiments by which M. Perreau established that these rays act on the electric resistance of selenium. New and valuable arguments have thus added force to those who incline towards a theory which has the merit of bringing a new phenomenon within the pale of phenomena previously known.

Nevertheless the shortest ultra-violet radiations, such as those of M. Schumann, are still capable of refraction by quartz, and this difference constitutes, in the minds of many physicists, a serious enough reason to decide them to reject the more simple hypothesis. Moreover, the rays of Schumann are, as we have seen, extraordinarily absorbable,—so much so that they have to be observed in a vacuum. The most striking property of the X rays is, on the contrary, the facility with which they pass through obstacles, and it is impossible not to attach considerable importance to such a difference.

Some attribute this marvellous radiation to longitudinal vibrations, which, as M. Duhem has shown, would be propagated in dielectric media with a speed equal to that of light. But the most generally accepted idea is the one formulated from the first by Sir George Stokes and followed up by Professor Wiechert. According to this theory the X rays should be due to a succession of independent pulsations of the ether, starting from the points where the molecules projected by the cathode of the Crookes tube meet the anticathode. These pulsations are not continuous vibrations like the radiations of the spectrum; they are isolated and extremely short; they are, besides, transverse, like the undulations of light, and the theory shows that they must be propagated with the speed of light. They should present neither refraction nor reflection, but, under certain conditions, they may be subject to the phenomena of diffraction. All these characteristics are found in the Röntgen rays.

Professor J.J. Thomson adopts an analogous idea, and states the precise way in which the pulsations may be produced at the moment when the electrified particles forming the cathode rays suddenly strike the anticathode wall. The electromagnetic induction behaves in such a way that the magnetic field is not annihilated when the particle stops, and the new field produced, which is no longer in equilibrium, is propagated in the dielectric like an electric pulsation. The electric and magnetic pulsations excited by this mechanism may give birth to effects similar to those of light. Their slight amplitude, however, is the cause of there here being neither refraction nor diffraction phenomena, save in very special conditions. If the cathode particle is not stopped in zero time, the pulsation will take a greater amplitude, and be, in consequence, more easily absorbable; to this is probably to be attributed the differences which may exist between different tubes and different rays.

It is right to add that some authors, notwithstanding the proved impossibility of deviating them in a magnetic field, have not renounced the idea of comparing them with the cathode rays. They suppose, for instance, that the rays are formed by electrons animated with so great a velocity that their inertia, conformably with theories which I shall examine later, no longer permit them to be stopped in their course; this is, for instance, the theory upheld by Mr Sutherland. We know, too, that to M. Gustave Le Bon they represent the extreme limit of material things, one of the last stages before the vanishing of matter on its return to the ether.

Everyone has heard of the N rays, whose name recalls the town of Nancy, where they were discovered. In some of their singular properties they are akin to the X rays, while in others they are widely divergent from them.

M. Blondlot, one of the masters of contemporary physics, deeply respected by all who know him, admired by everyone for the penetration of his mind, and the author of works remarkable for the originality and sureness of his method, discovered them in radiations emitted from various sources, such as the sun, an incandescent light, a Nernst lamp, and even bodies previously exposed to the sun's rays. The essential property which allows them to be revealed is their action on a small induction spark, of which they increase the brilliancy; this phenomenon is visible to the eye and is rendered objective by photography.

Various other physicists and numbers of physiologists, following the path opened by M. Blondlot, published during 1903 and 1904 manifold but often rather hasty memoirs, in which they related the results of their researches, which do not appear to have been always conducted with the accuracy desirable. These results were most strange; they seemed destined to revolutionise whole regions not only of the domain of physics, but likewise of the biological sciences. Unfortunately the method of observation was always founded on the variations in visibility of the spark or of a phosphorescent substance, and it soon became manifest that these variations were not perceptible to all eyes.

No foreign experimenter has succeeded in repeating the experiments, while in France many physicists have failed; and hence the question has much agitated public opinion. Are we face to face with a very singular case of suggestion, or is special training and particular dispositions required to make the phenomenon apparent? It is not possible, at the present moment, to declare the problem solved; but very recent experiments by M. Gutton and a note by M. Mascart have reanimated the confidence of those who hoped that such a scholar as M. Blondlot could not have been deluded by appearances. However, these last proofs in favour of the existence of the rays have themselves been contested, and have not succeeded in bringing conviction to everyone.

It seems very probable indeed that certain of the most singular conclusions arrived at by certain authors on the subject will lapse into deserved oblivion. But negative experiments prove nothing in a case like this, and the fact that most experimenters have failed where M. Blondlot and his pupils have succeeded may constitute a presumption, but cannot be regarded as a demonstrative argument. Hence we must still wait; it is exceedingly possible that the illustrious physicist of Nancy may succeed in discovering objective actions of the N rays which shall be indisputable, and may thus establish on a firm basis a discovery worthy of those others which have made his name so justly celebrated.

According to M. Blondlot the N rays can be polarised, refracted, and dispersed, while they have wavelengths comprised within .0030 micron, and .0760 micron—that is to say, between an eighth and a fifth of that found for the extreme ultra-violet rays. They might be, perhaps, simply rays of a very short period. Their existence, stripped of the parasitical and somewhat singular properties sought to be attributed to them, would thus appear natural enough. It would, moreover, be extremely important, and lead, no doubt, to most curious applications; it can be conceived, in fact, that such rays might serve to reveal what occurs in those portions of matter whose too minute dimensions escape microscopic examination on account of the phenomena of diffraction.

From whatever point of view we look at it, and whatever may be the fate of the discovery, the history of the N rays is particularly instructive, and must give food for reflection to those interested in questions of scientific methods.

§ 6. THE ETHER AND GRAVITATION

The striking success of the hypothesis of the ether in optics has, in our own days, strengthened the hope of being able to explain, by an analogous representation, the action of gravitation.

For a long time, philosophers who rejected the idea that ponderability is a primary and essential quality of all bodies have sought to reduce their weight to pressures exercised in a very subtle fluid. This was the conception of Descartes, and was perhaps the true idea of Newton himself. Newton points out, in many passages, that the laws he had discovered were independent of the hypotheses that could be formed on the way in which universal attraction was produced, but that with sufficient experiments the true cause of this attraction might one day be reached. In the preface to the second edition of the Optics he writes: "To prove that I have not considered weight as a universal property of bodies, I have added a question as to its cause, preferring this form of question because my interpretation does not entirely satisfy me in the absence of experiment"; and he puts the question in this shape: "Is not this medium (the ether) more rarefied in the interior of dense bodies like the sun, the planets, the comets, than in the empty spaces which separate them? Passing from these bodies to great distances, does it not become continually denser, and in that way does it not produce the weight of these great bodies with regard to each other and of their parts with regard to these bodies, each body tending to leave the most dense for the most rarefied parts?"

Evidently this view is incomplete, but we may endeavour to state it precisely. If we admit that this medium, the properties of which would explain the attraction, is the same as the luminous ether, we may first ask ourselves whether the action of gravitation is itself also due to oscillations. Some authors have endeavoured to found a theory on this hypothesis, but we are immediately brought face to face with very serious difficulties. Gravity appears, in fact, to present quite exceptional characteristics. No agent, not even those which depend upon the ether, such as light and electricity, has any influence on its action or its direction. All bodies are, so to speak, absolutely transparent to universal attraction, and no experiment has succeeded in demonstrating that its propagation is not instantaneous. From various astronomical observations, Laplace concluded that its velocity, in any case, must exceed fifty million times that of light. It is subject neither to reflection nor to refraction; it is independent of the structure of bodies; and not only is it inexhaustible, but also (as is pointed out, according to M. Hannequin, by an English scholar, James Croll) the distribution of the effects of the attracting force of a mass over the manifold particles which may successively enter the field of its action in no way diminishes the attraction it exercises on each of them respectively, a thing which is seen nowhere else in nature.

Nevertheless it is possible, by means of certain hypotheses, to construct interpretations whereby the appropriate movements of an elastic medium should explain the facts clearly enough. But these movements are very complex, and it seems almost inconceivable that the same medium could possess simultaneously the state of movement corresponding to the transmission of a luminous phenomenon and that constantly imposed on it by the transmission of gravitation.

Another celebrated hypothesis was devised by Lesage, of Geneva. Lesage supposed space to be overrun in all directions by currents of *ultramundane* corpuscles. This hypothesis, contested by Maxwell, is interesting. It might perhaps be taken up again in our days, and it is not impossible that the assimilation of these corpuscles to electrons might give a satisfactory image. [28]

M. Crémieux has recently undertaken experiments directed, as he thinks, to showing that the divergences between the phenomena of gravitation and all the other phenomena in nature are more apparent than real. Thus the evolution in the heart of the ether of a quantity of gravific energy would not be entirely isolated, and as in the case of all evolutions of all energy of whatever kind, it should provoke a partial transformation into energy of a different form. Thus again the liberated energy of gravitation would vary when passing from one material to another, as from gases into liquids, or from one liquid to a different one.

On this last point the researches of M. Crémieux have given affirmative results: if we immerse in a large mass of some liquid several drops of another not miscible with the first, but of identical density, we form a mass representing no doubt a discontinuity in the ether, and we may ask ourselves whether, in conformity with what happens in all other phenomena of nature, this discontinuity has not a tendency to disappear.

If we abide by the ordinary consequences of the Newtonian theory of potential, the drops should remain motionless, the hydrostatic impulsion forming an exact equilibrium to their mutual attraction. Now M. Crémieux remarks that, as a matter of fact, they slowly approach each other.

Such experiments are very delicate; and with all the precautions taken by the author, it cannot yet be asserted that he has removed all possibility of the action of the phenomena of capillarity nor all possible errors proceeding from extremely slight differences of temperature. But the attempt is interesting and deserves to be followed up.

Thus, the hypothesis of the ether does not yet explain all the phenomena which the considerations relating to matter are of themselves powerless to interpret. If we wished to represent to ourselves, by the mechanical properties of a medium filling the whole of the universe, all luminous, electric, and gravitation phenomena, we should be led to attribute to this medium very strange and almost contradictory characteristics; and yet it would be still more inconceivable that this medium should be double or treble, that there should be two or three ethers each occupying space as if it were alone, and interpenetrating it without exercising any action on one another. We are thus brought, by a close examination of facts, rather to the idea that the properties of the ether are not wholly reducible to the rules of ordinary mechanics.

The physicist has therefore not yet succeeded in answering the question often put to him by the philosopher: "Has the ether really an objective existence?" However, it is not necessary to know the answer in order to utilize the ether. In its ideal properties we find the means of determining the form of equations which are valid, and to the learned detached from all metaphysical prepossession this is the essential point.

CHAPTER VII
A CHAPTER IN THE HISTORY OF SCIENCE: WIRELESS TELEGRAPHY

§ 1

I have endeavoured in this book to set forth impartially the ideas dominant at this moment in the domain of physics, and to make known the facts essential to them. I have had to quote the authors of the principal discoveries in order to be able to class and, in some sort, to name these discoveries; but I in no way claim to write even a summary history of the physics of the day.

I am not unaware that, as has often been said, contemporary history is the most difficult of all histories to write. A certain step backwards seems necessary in order to enable us to appreciate correctly the relative importance of events, and details conceal the full view from eyes which are too close to them, as the trees prevent us from seeing the forest. The event which produces a great sensation has often only insignificant consequences; while another, which seemed at the outset of the least importance and little worthy of note, has in the long run a widespread and deep influence.

If, however, we deal with the history of a positive discovery, contemporaries who possess immediate information, and are in a position to collect authentic evidence at first hand, will make, by bringing to it their sincere testimony, a work of erudition which may be very useful, but which we may be tempted to look upon as very easy of execution. Yet such a labour, even when limited to the study of a very minute question or of a recent invention, is far from being accomplished without the historian stumbling over serious obstacles.

An invention is never, in reality, to be attributed to a single author. It is the result of the work of many collaborators who sometimes have no acquaintance with one another, and is often the fruit of obscure labours. Public opinion, however, wilfully simple in face of a sensational discovery, insists that the historian should also act as judge; and it is the historian's task to disentangle the truth in the midst of the contest, and to declare infallibly to whom the acknowledgments of mankind should be paid. He must, in his capacity as skilled expert, expose piracies, detect the most carefully hidden plagiarisms, and discuss the delicate question of priority; while he must not be deluded by those who do not fear to announce, in bold accents, that they have solved problems of which they find the solution imminent, and who, the day after its final elucidation by third parties, proclaim themselves its true discoverers. He must rise above a partiality which deems itself excusable because it proceeds from national pride; and, finally, he must seek with patience for what has gone before. While thus retreating step by step he runs the risk of losing himself in the night of time.

An example of yesterday seems to show the difficulties of such a task. Among recent discoveries the invention of wireless telegraphy is one of those which have rapidly become popular, and looks, as it were, an exact subject clearly marked out. Many attempts have already been made to write its history. Mr J.J. Fahie published in England as early as 1899 an interesting work entitled the *History of Wireless Telegraphy*; and about the same time M. Broca published in France a very exhaustive work named *La Telegraphie sans fil*. Among the reports presented to the Congrès international de physique (Paris, 1900), Signor Righi, an illustrious Italian scholar, whose personal efforts have largely contributed to the invention of the present system of telegraphy, devoted a chapter, short, but sufficiently complete, of his masterly report on Hertzian waves, to the history of wireless telegraphy. The same author, in association with Herr Bernhard Dessau, has likewise written a more important work, *Die Telegraphie ohne Draht*; and *La Telegraphie sans fil et les ondes Électriques* of MM. J. Boulanger and G. Ferrié may also be consulted with advantage, as may *La Telegraphie sans fil* of Signor Dominico Mazotto. Quite recently Mr A. Story has given us in a little volume called *The Story of Wireless Telegraphy*, a condensed but very precise recapitulation of all the attempts which have been made to establish telegraphic communication without the intermediary of a conducting wire. Mr Story has examined many documents, has sometimes brought curious facts to light, and has studied even the most recently adopted apparatus.

It may be interesting, by utilising the information supplied by these authors and supplementing them when necessary by others, to trace the sources of this modern discovery, to follow its developments, and thus to prove once more how much a matter, most simple in appearance, demands extensive and complex researches on the part of an author desirous of writing a definitive work.

§ 2

The first, and not the least difficulty, is to clearly define the subject. The words "wireless telegraphy," which at first seem to correspond to a simple and perfectly clear idea, may in reality apply to two series of questions, very different in the mind of a physicist, between which it is important to distinguish. The transmission of signals demands three organs which all appear indispensable: the transmitter, the receiver, and, between the two, an intermediary establishing the communication. This intermediary is generally the most costly part of the installation and the most difficult to set up, while it is here that the sensible losses of energy at the expense of good output occur. And yet our present ideas cause us to consider this intermediary as more than ever impossible to suppress; since, if we are definitely quit of the conception of action at a distance, it becomes inconceivable to us that energy can be communicated from one point to another without being carried by some intervening medium. But, practically, the line will be suppressed if, instead of constructing it artificially, we use to replace it one of the natural media which separate two points on the earth. These natural media are divided into two very distinct categories, and from this classification arise two series of questions to be examined.

Between the two points in question there are, first, the material media such as the air, the earth, and the water. For a long time we have used for transmissions to a distance the elastic properties of the air, and more recently the electric conductivity of the soil and of water, particularly that of the sea.

Modern physics leads us on the other hand, as we have seen, to consider that there exists throughout the whole of the universe another and more subtle medium which penetrates everywhere, is endowed with elasticity *in vacuo*, and retains its elasticity when it penetrates into a great number of bodies, such as the air. This medium is the luminous ether which possesses, as we cannot doubt, the property of being able to transmit energy, since it itself brings to us by far the larger part of the energy which we possess on earth and which we find in the movements of the atmosphere, or of waterfalls, and in the coal mines proceeding from the decomposition of carbon compounds under the influence of the solar energy. For a long time also before the existence of the ether was known, the duty of transmitting signals was entrusted to it. Thus through the ages a double evolution is unfolded which has to be followed by the historian who is ambitious of completeness.

§ 3

If such an historian were to examine from the beginning the first order of questions, he might, no doubt, speak only briefly of the attempts earlier than electric telegraphy. Without seeking to be paradoxical, he certainly ought to mention the invention of the speaking-trumpet and other similar inventions which for a long time have enabled mankind, by the ingenious use of the elastic properties of the natural media, to communicate at greater distances than they could have attained without the aid of art. After this in some sort prehistoric period had been rapidly run through, he would have to follow very closely the development of electric telegraphy. Almost from the outset, and shortly after Ampère had made public the idea of constructing a telegraph, and the day after Gauss and Weber set up between their houses in Göttingen the first line really used, it was thought that the conducting properties of the earth and water might be made of service.

The history of these trials is very long, and is closely mixed up with the history of ordinary telegraphy; long chapters for some time past have been devoted to it in telegraphic treatises. It was in 1838, however, that Professor C.A. Steinheil of Munich expressed, for the first time, the clear idea of suppressing the return wire and replacing it by a connection of the line wire to the earth. He thus at one step covered half the way, the easiest, it is true, which was to lead to the final goal, since he saved the use of one-half of the line of wire. Steinheil, advised, perhaps, by Gauss, had, moreover, a very exact conception of the part taken by the earth considered as a conducting body. He seems to have well understood that, in certain conditions, the resistance of such a conductor, though supposed to be unlimited, might be independent of the distance apart of the electrodes which carry the current and allow it to go forth. He likewise thought of using the railway lines to transmit telegraphic signals.

Several scholars who from the first had turned their minds to telegraphy, had analogous ideas. It was thus that S.F.B. Morse, superintendent of the Government telegraphs in the United States, whose name is universally known in connection with the very simple apparatus invented by him, made experiments in the autumn of 1842 before a special commission in New York and a numerous public audience, to show how surely and how easily his apparatus worked. In the very midst of his experiments a very happy idea occurred to him of replacing by the water of a canal, the length of about a mile of wire which had been suddenly and accidentally destroyed. This accident, which for a moment compromised the legitimate success the celebrated engineer expected, thus suggested to him a fruitful idea which he did not forget. He subsequently repeated attempts to thus utilise the earth and water, and obtained some very remarkable results.

It is not possible to quote here all the researches undertaken with the same purpose, to which are more particularly attached the names of S.W. Wilkins, Wheatstone, and H. Highton, in England; of Bonetti in Italy, Gintl in Austria,

Bouchot and Donat in France; but there are some which cannot be recalled without emotion.

On the 17th December 1870, a physicist who has left in the University of Paris a lasting name, M. d'Almeida, at that time Professor at the Lycée Henri IV. and later Inspector-General of Public Instruction, quitted Paris, then besieged, in a balloon, and descended in the midst of the German lines. He succeeded, after a perilous journey, in gaining Havre by way of Bordeaux and Lyons; and after procuring the necessary apparatus in England, he descended the Seine as far as Poissy, which he reached on the 14th January 1871. After his departure, two other scholars, MM. Desains and Bourbouze, relieving each other day and night, waited at Paris, in a wherry on the Seine, ready to receive the signal which they awaited with patriotic anxiety. It was a question of working a process devised by the last-named pair, in which the water of the river acted the part of the line wire. On the 23rd January the communication at last seemed to be established, but unfortunately, first the armistice and then the surrender of Paris rendered useless the valuable result of this noble effort.

Special mention is also due to the experiments made by the Indian Telegraph Office, under the direction of Mr Johnson and afterwards of Mr W.F. Melhuish. They led, indeed, in 1889 to such satisfactory results that a telegraph service, in which the line wire was replaced by the earth, worked practically and regularly. Other attempts were also made during the latter half of the nineteenth century to transmit signals through the sea. They preceded the epoch when, thanks to numerous physicists, among whom Lord Kelvin undoubtedly occupies a preponderating position, we succeeded in sinking the first cable; but they were not abandoned, even after that date, for they gave hopes of a much more economical solution of the problem. Among the most interesting are remembered those that S.W. Wilkins carried on for a long time between France and England. Like Cooke and Wheatstone, he thought of using as a receiver an apparatus which in some features resembles the present receiver of the submarine telegraph. Later, George E. Dering, then James Bowman and Lindsay, made on the same lines trials which are worthy of being remembered.

But it is only in our own days that Sir William H. Preece at last obtained for the first time really practical results. Sir William himself effected and caused to be executed by his associates—he is chief consulting engineer to the General Post Office in England—researches conducted with much method and based on precise theoretical considerations. He thus succeeded in establishing very easy, clear, and regular communications between various places; for example, across the Bristol Channel. The long series of operations accomplished by so many seekers, with the object of substituting a material and natural medium for the artificial lines of metal, thus met with an undoubted success which was soon to be eclipsed by the widely-known experiments directed into a different line by Marconi.

It is right to add that Sir William Preece had himself utilised induction phenomena in his experiments, and had begun researches with the aid of electric waves. Much is due to him for the welcome he gave to Marconi; it is certainly thanks to the advice and the material support he found in Sir William that the young scholar succeeded in effecting his sensational experiments.

§ 4

The starting-point of the experiments based on the properties of the luminous ether, and having for their object the transmission of signals, is very remote; and it would be a very laborious task to hunt up all the work accomplished in that direction, even if we were to confine ourselves to those in which electrical reactions play a part. An electric reaction, an electrostatic influence, or an electromagnetic phenomenon, is transmitted at a distance through the air by the intermediary of the luminous ether. But electric influence can hardly be used, as the distances it would allow us to traverse would be much too restricted, and electrostatic actions are often very erratic. The phenomena of induction, which are very regular and insensible to the variations of the atmosphere, have, on the other hand, for a long time appeared serviceable for telegraphic purposes.

We might find, in a certain number of the attempts just mentioned, a partial employment of these phenomena. Lindsay, for instance, in his project of communication across the sea, attributed to them a considerable rôle. These phenomena even permitted a true telegraphy without intermediary wire between the transmitter and the receiver, at very restricted distances, it is true, but in peculiarly interesting conditions. It is, in fact, owing to them that C. Brown, and later Edison and Gilliland, succeeded in establishing communications with trains in motion.

Mr Willoughby S. Smith and Mr Charles A. Stevenson also undertook experiments during the last twenty years, in which they used induction, but the most remarkable attempts are perhaps those of Professor Emile Rathenau. With the assistance of Professor Rubens and of Herr W. Rathenau, this physicist effected, at the request of the German Ministry of Marine, a series of researches which enabled him, by means of a compound system of conduction and induction by alternating currents, to obtain clear and regular communications at a distance of four kilometres. Among the precursors also should be mentioned Graham Bell; the inventor of the telephone thought of employing his admirable apparatus as a receiver of induction phenomena transmitted from a distance; Edison, Herr Sacher of Vienna, M. Henry Dufour of Lausanne, and Professor Trowbridge of Boston, also made interesting attempts in the same direction.

In all these experiments occurs the idea of employing an oscillating current. Moreover, it was known for a long time—since, in 1842, the great American physicist Henry proved that the discharges from a Leyden jar in the attic of his house caused sparks in a metallic circuit on the ground floor— that a flux which varies rapidly and periodically is much more efficacious than a simple flux, which latter can only produce at a distance a phenomenon of slight intensity. This idea of the oscillating current was closely akin to that which was at last to lead to an entirely satisfactory solution: that is, to a solution which is founded on the properties of electric waves.

§ 5

Having thus got to the threshold of the definitive edifice, the historian, who has conducted his readers over the two parallel routes which have just been marked out, will be brought to ask himself whether he has been a sufficiently faithful guide and has not omitted to draw attention to all essential points in the regions passed through.

Ought we not to place by the side, or perhaps in front, of the authors who have devised the practical appliances, those scholars who have constructed the theories and realised the laboratory experiments of which, after all, the apparatus are only the immediate applications? If we speak of the propagation of a current in a material medium, can one forget the names of Fourier and of Ohm, who established by theoretical considerations the laws which preside over this propagation? When one looks at the phenomena of induction, would it not be just to remember that Arago foresaw them, and that Michael Faraday discovered them? It would be a delicate, and also a rather puerile task, to class men of genius in order of merit. The merit of an inventor like Edison and that of a theorist like Clerk Maxwell have no common measure, and mankind is indebted for its great progress to the one as much as to the other.

Before relating how success attended the efforts to utilise electric waves for the transmission of signals, we cannot without ingratitude pass over in silence the theoretical speculations and the work of pure science which led to the knowledge of these waves. It would therefore be just, without going further back than Faraday, to say how that illustrious physicist drew attention to the part taken by insulating media in electrical phenomena, and to insist also on the admirable memoirs in which for the first time Clerk Maxwell made a solid bridge between those two great chapters of Physics, optics and electricity, which till then had been independent of each other. And no doubt it would be impossible not to evoke the memory of those who, by establishing, on the other hand, the solid and magnificent structure of physical optics, and proving by their immortal works the undulatory nature of light, prepared from the opposite direction the future unity. In the history of the applications of electrical undulations, the names of Young, Fresnel, Fizeau, and Foucault must be inscribed; without these scholars, the assimilation between electrical and luminous phenomena which they discovered and studied would evidently have been impossible.

Since there is an absolute identity of nature between the electric and the luminous waves, we should, in all justice, also consider as precursors those who devised the first luminous telegraphs. Claude Chappe incontestably effected wireless telegraphy, thanks to the luminous ether, and the learned men, such as Colonel Mangin, who perfected optical telegraphy, indirectly suggested certain improvements lately introduced into the present method.

But the physicist whose work should most of all be put in evidence is, without fear of contradiction, Heinrich Hertz. It was he who demonstrated irrefutably, by experiments now classic, that an electric discharge produces an undulatory disturbance in the ether contained in the insulating media in its neighbourhood; it was he who, as a profound theorist, a clever mathematician, and an experimenter of prodigious dexterity, made known the mechanism of the production, and fully elucidated that of the propagation of these electromagnetic waves.

He must naturally himself have thought that his discoveries might be applied to the transmission of signals. It would appear, however, that when interrogated by a Munich engineer named Huber as to the possibility of utilising the waves for transmissions by telephone, he answered in the negative, and dwelt on certain considerations relative to the difference between the periods of sounds and those of electrical vibrations. This answer does not allow us to judge what might have happened, had not a cruel death carried off in 1894, at the age of thirty-five, the great and unfortunate physicist.

We might also find in certain works earlier than the experiments of Hertz attempts at transmission in which, unconsciously no doubt, phenomena were already set in operation which would, at this day, be classed as electric oscillations. It is allowable no doubt, not to speak of an American quack, Mahlon Loomis, who, according to Mr Story, patented in 1870 a project of communication in which he utilised the Rocky Mountains on one side and Mont Blanc on the other, as gigantic antennae to establish communication across the Atlantic; but we cannot pass over in silence the very remarkable researches of the American Professor Dolbear, who showed, at the electrical exhibition of Philadelphia in 1884, a set of apparatus enabling signals to be transmitted at a distance, which he described as "an exceptional application of the principles of electrostatic induction." This apparatus comprised groups of coils and condensers by means of which he obtained, as we cannot now doubt, effects due to true electric waves.

Place should also be made for a well-known inventor, D.E. Hughes, who from 1879 to 1886 followed up some very curious experiments in which also these oscillations certainly played a considerable part. It was this physicist who invented the microphone, and thus, in another way, drew attention to the variations of contact resistance, a phenomenon not far from that produced in the radio-conductors of Branly, which are important organs in the Marconi system. Unfortunately, fatigued and in ill-health, Hughes ceased his researches at the moment perhaps when they would have given him final results.

In an order of ideas different in appearance, but closely linked at bottom with the one just mentioned, must be recalled the discovery of radiophony in 1880 by Graham Bell, which was foreshadowed in 1875 by C.A. Brown. A luminous ray falling on a selenium cell produces a variation of electric resistance, thanks to which a sound signal can be transmitted by light. That delicate instrument the radiophone, constructed on this principle, has wide analogies with the apparatus of to-day.

§ 6

Starting from the experiments of Hertz, the history of wireless telegraphy almost merges into that of the researches on electrical waves. All the progress realised in the manner of producing and receiving these waves necessarily helped to give rise to the application already indicated. The experiments of Hertz, after being checked in every laboratory, and having entered into the strong domain of our most certain knowledge, were about to yield the expected fruit.

Experimenters like Sir Oliver Lodge in England, Righi in Italy, Sarrazin and de la Rive in Switzerland, Blondlot in France, Lecher in Germany, Bose in India, Lebedeff in Russia, and theorists like M.H. Poincaré and Professor Bjerknes, who devised ingenious arrangements or elucidated certain points left dark, are among the artisans of the work which followed its natural evolution.

It was Professor R. Threlfall who seems to have been the first to clearly propose, in 1890, the application of the Hertzian waves to telegraphy, but it was certainly Sir W. Crookes who, in a very remarkable article in the *Fortnightly Review* of February 1892, pointed out very clearly the road to be followed. He even showed in what conditions the Morse receiver might be applied to the new system of telegraphy.

About the same period an American physicist, well known by his celebrated experiments on high frequency currents—experiments, too, which are not unconnected with those on electric oscillations,—M. Tesla, demonstrated that these oscillations could be transmitted to more considerable distances by making use of two vertical antennae, terminated by large conductors.

A little later, Sir Oliver Lodge succeeded, by the aid of the coherer, in detecting waves at relatively long distances, and Mr Rutherford obtained similar results with a magnetic indicator of his own invention.

An important question of meteorology, the study of atmospheric discharges, at this date led a few scholars, and more particularly the Russian, M. Popoff, to set up apparatus very analogous to the receiving apparatus of the present wireless telegraphy. This comprised a long antenna and filings-tube, and M. Popoff even pointed out that his apparatus might well serve for the transmission of signals as soon as a generator of waves powerful enough had been discovered.

Finally, on the 2nd June 1896, a young Italian, born in Bologna on the 25th April 1874, Guglielmo Marconi, patented a system of wireless telegraphy destined to become rapidly popular. Brought up in the laboratory of Professor Righi, one of the physicists who had done most to confirm and extend the experiments of Hertz, Marconi had long been familiar with the properties of electric waves, and was well used to their manipulation. He afterwards had the good fortune to meet Sir William (then Mr) Preece, who was to him an adviser of the highest authority.

It has sometimes been said that the Marconi system contains nothing original; that the apparatus for producing the waves was the oscillator of Righi, that the receiver was that employed for some two or three years by Professor Lodge and Mr Bose, and was founded on an earlier discovery by a French scholar, M. Branly; and, finally, that the general arrangement was that established by M. Popoff.

The persons who thus rather summarily judge the work of M. Marconi show a severity approaching injustice. It cannot, in truth, be denied that the young scholar has brought a strictly personal contribution to the solution of the problem he proposed to himself. Apart from his forerunners, and when their attempts were almost unknown, he had the very great merit of adroitly arranging the most favourable combination, and he was the first to succeed in obtaining practical results, while he showed that the electric waves could be transmitted and received at distances enormous compared to those attained before his day. Alluding to a well-known anecdote relating to Christopher Columbus, Sir W. Preece very justly said: "The forerunners and rivals of Marconi no doubt knew of the eggs, but he it was who taught them to make them stand on end." This judgment will, without any doubt, be the one that history will definitely pronounce on the Italian scholar.

§ 7

The apparatus which enables the electric waves to be revealed, the detector or indicator, is the most delicate organ in wireless telegraphy. It is not necessary to employ as an indicator a filings-tube or radio-conductor. One can, in principle, for the purpose of constructing a receiver, think of any one of the multiple effects produced by the Hertzian waves. In many systems in use, and in the new one of Marconi himself, the use of these tubes has been abandoned and replaced by magnetic detectors.

Nevertheless, the first and the still most frequent successes are due to radio-conductors, and public opinion has not erred in attributing to the inventor of this ingenious apparatus a considerable and almost preponderant part in the invention of wave telegraphy.

The history of the discovery of radio-conductors is short, but it deserves, from its importance, a chapter to itself in the history of wireless telegraphy. From a theoretical point of view, the phenomena produced in those tubes should be set by the side of those studied by Graham Bell, C.A. Brown, and Summer Tainter, from the year 1878 onward. The variations to which luminous waves give rise in the resistance of selenium and other substances are, doubtless, not unconnected with those which the electric waves produce in filings. A connection can also be established between this effect of the waves and the variations of contact resistance which enabled Hughes to construct the microphone, that admirable instrument which is one of the essential organs of telephony.

More directly, as an antecedent to the discovery, should be quoted the remark made by Varley in 1870, that coal-dust changes in conductivity when the electromotive force of the current which passes through it is made to vary. But it was in 1884 that an Italian professor, Signor Calzecchi-Onesti, demonstrated in a series of remarkable experiments that the metallic filings contained in a tube of insulating material, into which two metallic electrodes are inserted, acquire a notable conductivity under different influences such as extra currents, induced currents, sonorous vibrations, etc., and that this conductivity is easily destroyed; as, for instance, by turning the tube over and over.

In several memoirs published in 1890 and 1891, M. Ed. Branly independently pointed out similar phenomena, and made a much more complete and systematic study of the question. He was the first to note very clearly that the action described could be obtained by simply making sparks pass in the neighbourhood of the radio-conductor, and that their great resistance could be restored to the filings by giving a slight shake to the tube or to its supports.

The idea of utilising such a very interesting phenomenon as an indicator in the study of the Hertzian waves seems to have occurred simultaneously to several physicists, among whom should be especially mentioned M. Ed. Branly himself, Sir Oliver Lodge, and MM. Le Royer and Van Beschem, and its use in laboratories rapidly became quite common.

The action of the waves on metallic powders has, however, remained some what mysterious; for ten years it has been the subject of important researches by Professor Lodge, M. Branly, and a very great number of the most distinguished physicists. It is impossible to notice here all these researches, but from a recent and very interesting work of M. Blanc, it would seem that the phenomenon is allied to that of ionisation.

§ 8

The history of wireless telegraphy does not end with the first experiments of Marconi; but from the moment their success was announced in the public press, the question left the domain of pure science to enter into that of commerce. The historian's task here becomes different, but even more delicate; and he will encounter difficulties which can be only known to one about to write the history of a commercial invention.

The actual improvements effected in the system are kept secret by the rival companies, and the most important results are patriotically left in darkness by the learned officers who operate discreetly in view of the national defence. Meanwhile, men of business desirous of bringing out a company proclaim, with great nourish of advertisements, that they are about to exploit a process superior to all others.

On this slippery ground the impartial historian must nevertheless venture; and he may not refuse to relate the progress accomplished, which is considerable. Therefore, after having described the experiments carried out for nearly ten years by Marconi himself, first across the Bristol Channel, then at Spezzia, between the coast and the ironclad *San Bartolommeo*, and finally by means of gigantic apparatus between America and England, he must give the names of those who, in the different civilised countries, have contributed to the improvement of the system of communication by waves; while he must describe what precious services this system has already rendered to the art of war, and happily also to peaceful navigation.

From the point of view of the theory of the phenomena, very remarkable results have been obtained by various physicists, among whom should be particularly mentioned M. Tissot, whose brilliant studies have thrown a bright light on different interesting points, such as the rôle of the antennae. It would be equally impossible to pass over in silence other recent attempts in a slightly different groove. Marconi's system, however improved it may be to-day, has one grave defect. The synchronism of the two pieces of apparatus, the transmitter and the receiver, is not perfect, so that a message sent off by one station may be captured by some other station. The fact that the phenomena of resonance are not utilised, further prevents the quantity of energy received by the receiver from being considerable, and hence the effects reaped are very weak, so that the system remains somewhat fitful and the communications are often disturbed by atmospheric phenomena. Causes which render the air a momentary conductor, such as electrical discharges, ionisation, etc., moreover naturally prevent the waves from passing, the ether thus losing its elasticity.

Professor Ferdinand Braun of Strasburg has conceived the idea of employing a mixed system, in which the earth and the water, which, as we have seen, have often been utilised to conduct a current for transmitting a signal, will serve as a sort of guide to the waves themselves. The now well-known theory of the propagation of waves guided by a conductor enables it to be foreseen that, according to their periods, these waves will penetrate more or less deeply into the natural medium, from which fact has been devised a method of separating them according to their frequency. By applying this theory, M. Braun has carried out, first in the fortifications of Strasburg, and then between the island of Heligoland and the mainland, experiments which have given remarkable results. We might mention also the researches, in a somewhat analogous order of ideas, by an English engineer, Mr Armstrong, by Dr Lee de Forest, and also by Professor Fessenden.

Having thus arrived at the end of this long journey, which has taken him from the first attempts down to the most recent experiments, the historian can yet set up no other claim but that of having written the commencement of a history which others must continue in the future. Progress does not stop, and it is never permissible to say that an invention has reached its final form.

Should the historian desire to give a conclusion to his labour and answer the question the reader would doubtless not fail to put to him, "To whom, in short, should the invention of wireless telegraphy more particularly be attributed?" he should certainly first give the name of Hertz, the genius who discovered the waves, then that of Marconi, who was the first to transmit signals by the use of Hertzian undulations, and should add those of the scholars who, like Morse, Popoff, Sir W. Preece, Lodge, and, above all, Branly, have devised the arrangements necessary for their transmission. But he might then recall what Voltaire wrote in the *Philosophical Dictionary*:

"What! We wish to know what was the exact theology of Thot, of Zerdust, of Sanchuniathon, of the first Brahmins, and we are ignorant of the inventor of the shuttle! The first weaver, the first mason, the first smith, were no doubt great geniuses, but they were disregarded. Why? Because none of them invented a perfected art. The one who hollowed out an oak to cross a river never made a galley; those who piled up rough stones with girders of wood did not plan the Pyramids. Everything is made by degrees and the glory belongs to no one."

To-day, more than ever, the words of Voltaire are true: science becomes more and more impersonal, and she teaches us that progress is nearly always due to the united efforts of a crowd of workers, and is thus the best school of social solidarity.

CHAPTER VIII
THE CONDUCTIVITY OF GASES AND THE IONS
§ 1. THE CONDUCTIVITY OF GASES

If we were confined to the facts I have set forth above, we might conclude that two classes of phenomena are to-day being interpreted with increasing correctness in spite of the few difficulties which have been pointed out. The hypothesis of the molecular constitution of matter enables us to group together one of these classes, and the hypothesis of the ether leads us to co-ordinate the other.

But these two classes of phenomena cannot be considered independent of each other. Relations evidently exist between matter and the ether, which manifest themselves in many cases accessible to experiment, and the search for these relations appears to be the paramount problem the physicist should set himself. The question has, for a long time, been attacked on various sides, but the recent discoveries in the conductivity of gases, of the radioactive substances, and of the cathode and similar rays, have allowed us of late years to regard it in a new light. Without wishing to set out here in detail facts which for the most part are well known, we will endeavour to group the chief of them round a few essential ideas, and will seek to state precisely the data they afford us for the solution of this grave problem.

It was the study of the conductivity of gases which at the very first furnished the most important information, and allowed us to penetrate more deeply than had till then been possible into the inmost constitution of matter, and thus to, as it were, catch in the act the actions that matter can exercise on the ether, or, reciprocally, those it may receive from it.

It might, perhaps, have been foreseen that such a study would prove remarkably fruitful. The examination of the phenomena of electrolysis had, in fact, led to results of the highest importance on the constitution of liquids, and the gaseous media which presented themselves as particularly simple in all their properties ought, it would seem, to have supplied from the very first a field of investigation easy to work and highly productive.

This, however, was not at all the case. Experimental complications springing up at every step obscured the problem. One generally found one's self in the presence of violent disruptive discharges with a train of accessory phenomena, due, for instance, to the use of metallic electrodes, and made evident by the complex appearance of aigrettes and effluves; or else one had to deal with heated gases difficult to handle, which were confined in receptacles whose walls played a troublesome part and succeeded in veiling the simplicity of the fundamental facts. Notwithstanding, therefore, the efforts of a great number of seekers, no general idea disengaged itself out of a mass of often contradictory information.

Many physicists, in France particularly, discarded the study of questions which seemed so confused, and it must even be frankly acknowledged that some among them had a really unfounded distrust of certain results which should have been considered proved, but which had the misfortune to be in contradiction with the theories in current use. All the classic ideas relating to electrical phenomena led to the consideration that there existed a perfect symmetry between the two electricities, positive and negative. In the passing of electricity through gases there is manifested, on the contrary, an evident dissymmetry. The anode and the cathode are immediately distinguished in a tube of rarefied gas by their peculiar appearance; and the conductivity does not appear, under certain conditions, to be the same for the two modes of electrification.

It is not devoid of interest to note that Erman, a German scholar, once very celebrated and now generally forgotten, drew attention as early as 1815 to the unipolar conductivity of a flame. His contemporaries, as may be gathered from the perusal of the treatises on physics of that period, attached great importance to this discovery; but, as it was somewhat inconvenient and did not readily fit in with ordinary studies, it was in due course neglected, then considered as insufficiently established, and finally wholly forgotten.

All these somewhat obscure facts, and some others—such as the different action of ultra-violet radiations on positively and negatively charged bodies—are now, on the contrary, about to be co-ordinated, thanks to the modern ideas on the mechanism of conduction; while these ideas will also allow us to interpret the most striking dissymmetry of all, *i.e.* that revealed by electrolysis itself, a dissymmetry which certainly can not be denied, but to which sufficient attention has not been given.

It is to a German physicist, Giese, that we owe the first notions on the mechanism of the conductivity of gases, as we now conceive it. In two memoirs published in 1882 and 1889, he plainly arrives at the conception that conduction in gases is not due to their molecules, but to certain fragments of them or to ions. Giese was a forerunner, but his ideas could not triumph so long as there were no means of observing conduction in simple circumstances. But this means has now been supplied in the discovery of the X rays. Suppose we pass through some gas at ordinary pressure, such as hydrogen, a pencil of X rays. The gas, which till then has behaved as a perfect insulator,[29] suddenly acquires a remarkable conductivity. If into this hydrogen two metallic electrodes in communication with the two poles of a battery are introduced, a current is set up in very special conditions which remind us, when they are checked by experiments, of the mechanism which allows the passage of electricity in electrolysis, and which is so well represented to us when we picture to ourselves this passage as due to the migration towards the electrodes, under the action of the field, of the two sets of ions produced by the spontaneous division of the molecule within the solution.

Let us therefore recognise with J.J. Thomson and the many physicists who, in his wake, have taken up and developed the idea of Giese, that, under the influence of the X rays, for reasons which will have to be determined later, certain gaseous molecules have become divided into two portions, the one positively and the other negatively electrified, which we will call, by analogy with the kindred phenomenon in electrolysis, by the name of ions. If the gas be then placed in an electric field, produced, for instance, by two metallic plates connected with the two poles of a battery respectively, the positive ions will travel towards the plate connected with the negative pole, and the negative ions in the contrary direction. There is thus produced a current due to the transport to the electrodes of the charges which existed on the ions.

If the gas thus ionised be left to itself, in the absence of any electric field, the ions, yielding to their mutual attraction, must finally meet, combine, and reconstitute a neutral molecule, thus returning to their initial condition. The gas in a short while loses the conductivity which it had acquired; or this is, at least, the phenomenon at ordinary temperatures. But if the temperature is raised, the relative speeds of the ions at the moment of impact may be great enough to render it impossible for the recombination to be produced in its entirety, and part of the conductivity will remain.

Every element of volume rendered a conductor therefore furnishes, in an electric field, equal quantities of positive and negative electricity. If we admit, as mentioned above, that these liberated quantities are borne by ions each bearing an equal charge, the number of these ions will be proportional to the quantity of electricity, and instead of speaking of a quantity of electricity, we could use the equivalent term of number of ions. For the excitement produced by a given pencil of X rays, the number of ions liberated will be fixed. Thus, from a given volume of gas there can only be extracted an equally determinate quantity of electricity.

The conductivity produced is not governed by Ohm's law. The intensity is not proportional to the electromotive force, and it increases at first as the electromotive force augments; but it approaches asymptotically to a maximum value which corresponds to the number of ions liberated, and can therefore serve as a measure of the power of the excitement. It is this current which is termed the *current of saturation*.

M. Righi has ably demonstrated that ionised gas does not obey the law of Ohm by an experiment very paradoxical in appearance. He found that, the greater the distance of the two electrode plates from each, the greater may be, within certain limits, the intensity of the current. The fact is very clearly interpreted by the theory of ionisation, since the greater the length of the gaseous column the greater must be the number of ions liberated.

One of the most striking characteristics of ionised gases is that of discharging electrified conductors. This phenomenon is not produced by the departure of the charge that these conductors may possess, but by the advent of opposite charges brought to them by ions which obey the electrostatic attraction and abandon their own electrification when they come in contact with these conductors.

This mode of regarding the phenomena is extremely convenient and eminently suggestive. It may, no doubt, be thought that the image of the ions is not identical with objective reality, but we are compelled to acknowledge that it represents with absolute faithfulness all the details of the phenomena.

Other facts, moreover, will give to this hypothesis a still greater value; we shall even be able, so to speak, to grasp these ions individually, to count them, and to measure their charge.

§ 2. THE CONDENSATION OF WATER-VAPOUR BY IONS

If the pressure of a vapour—that of water, for instance—in the atmosphere reaches the value of the maximum pressure corresponding to the temperature of the experiment, the elementary theory teaches us that the slightest decrease in temperature will induce a condensation; that small drops will form, and the mist will turn into rain.

In reality, matters do not occur in so simple a manner. A more or less considerable delay may take place, and the vapour will remain supersaturated. We easily discover that this phenomenon is due to the intervention of capillary action. On a drop of liquid a surface-tension takes effect which gives rise to a pressure which becomes greater the smaller the diameter of the drop.

Pressure facilitates evaporation, and on more closely examining this reaction we arrive at the conclusion that vapour can never spontaneously condense itself when liquid drops already formed are not present, unless forces of another nature intervene to diminish the effect of the capillary forces. In the most frequent cases, these forces come from the dust which is always in suspension in the air, or which exists in any recipient. Grains of dust act by reason of their hygrometrical power, and form germs round which drops presently form. It is possible to make use, as did M. Coulier as early as 1875, of this phenomenon to carry off the germs of condensation, by producing by expansion in a bottle containing a little water a preliminary mist which purifies the air. In subsequent experiments it will be found almost impossible to produce further condensation of vapour.

But these forces may also be of electrical origin. Von Helmholtz long since showed that electricity exercises an influence on the condensation of the vapour of water, and Mr C.T.R. Wilson, with this view, has made truly quantitative experiments. It was rapidly discovered after the apparition of the X rays that gases that have become conductors, that is, ionised gases, also facilitate the condensation of supersaturated water vapour.

We are thus led by a new road to the belief that electrified centres exist in gases, and that each centre draws to itself the neighbouring molecules of water, as an electrified rod of resin does the light bodies around it. There is produced in this manner round each ion an assemblage of molecules of water which constitute a germ capable of causing the formation of a drop of water out of the condensation of excess vapour in the ambient air. As might be expected, the drops are electrified, and take to themselves the charge of the centres round which they are formed; moreover, as many drops are created as there are ions. Thereafter we have only to count these drops to ascertain the number of ions which existed in the gaseous mass.

To effect this counting, several methods have been used, differing in principle but leading to similar results. It is possible, as Mr C.T.R. Wilson and Professor J.J. Thomson have done, to estimate, on the one hand, the weight of the mist which is produced in determined conditions, and on the other, the average weight of the drops, according to the formula formerly given by Sir G. Stokes, by deducting their diameter from the speed with which this mist falls; or we can, with Professor Lemme, determine the average radius of the drops by an optical process, viz. by measuring the diameter of the first diffraction ring produced when looking through the mist at a point of light.

We thus get to a very high number. There are, for instance, some twenty million ions per centimetre cube when the rays have produced their maximum effect, but high as this figure is, it is still very small compared with the total number of molecules. All conclusions drawn from kinetic theory lead us to think that in the same space there must exist, by the side of a molecule divided into two ions, a thousand millions remaining in a neutral state and intact.

Mr C.T.R. Wilson has remarked that the positive and negative ions do not produce condensation with the same facility. The ions of a contrary sign may be almost completely separated by placing the ionised gas in a suitably disposed field. In the neighbourhood of a negative disk there remain hardly any but positive ions, and against a positive disk none but negative; and in effecting a separation of this kind, it will be noticed that condensation by negative ions is easier than by the positive.

It is, consequently, possible to cause condensation on negative centres only, and to study separately the phenomena produced by the two kinds of ions. It can thus be verified that they really bear charges equal in absolute value, and these charges can even be estimated, since we already know the number of drops. This estimate can be made, for example, by comparing the speed of the fall of a mist in fields of different values, or, as did J.J. Thomson, by measuring the total quantity of electricity liberated throughout the gas.

At the degree of approximation which such experiments imply, we find that the charge of a drop, and consequently the charge borne by an ion, is sensibly 3.4×10^{-10} electrostatic or 1.1×10^{-20} electromagnetic units. This charge is very near that which the study of the phenomena of ordinary electrolysis leads us to attribute to a univalent atom produced by electrolytic dissociation.

Such a coincidence is evidently very striking; but it will not be the only one, for whatever phenomenon be studied it will always appear that the smallest charge we can conceive as isolated is that mentioned. We are, in fact, in presence of a natural unit, or, if you will, of an atom of electricity.

We must, however, guard against the belief that the gaseous ion is identical with the electrolytic ion. Sensible differences between those are immediately apparent, and still greater ones will be discovered on closer examination.

As M. Perrin has shown, the ionisation produced by the X-rays in no way depends on the chemical composition of the gas; and whether we take a volume of gaseous hydrochloric acid or a mixture of hydrogen and chlorine in the same condition, all the results will be identical: and chemical affinities play no part here.

We can also obtain other information regarding ions: we can ascertain, for instance, their velocities, and also get an idea of their order of magnitude.

By treating the speeds possessed by the liberated charges as components of the known speed of a gaseous current, Mr Zeleny measures the mobilities, that is to say, the speeds acquired by the positive and negative charges in a field equal to the electrostatic unit. He has thus found that these mobilities are different, and that they vary, for example, between 400 and 200 centimetres per second for the two charges in dry gases, the positive being less mobile than the negative ions, which suggests the idea that they are of greater mass.[30]

M. Langevin, who has made himself the eloquent apostle of the new doctrines in France, and has done much to make them understood and admitted, has personally undertaken experiments analogous to those of M. Zeleny, but much more complete. He has studied in a very ingenious manner, not only the mobilities, but also the law of recombination which regulates the spontaneous return of the gas to its normal state. He has determined experimentally the relation of the number of recombinations to the number of collisions between two ions of contrary sign, by studying the variation produced by a change in the value of the field, in the quantity of electricity which can be collected in the gas separating two parallel metallic plates, after the passage through it for a very short time of the Röntgen rays emitted during one discharge of a Crookes tube. If the image of the ions is indeed conformable to reality, this relation must evidently always be smaller than unity, and must tend towards this value when the mobility of the ions diminishes, that is to say, when the pressure of the gas increases. The results obtained are in perfect accord with this anticipation.

On the other hand, M. Langevin has succeeded, by following the displacement of the ions between the parallel plates after the ionisation produced by the radiation, in determining the absolute values of the mobilities with great precision, and has thus clearly placed in evidence the irregularity of the mobilities of the positive and negative ions respectively. Their mass can be calculated when we know, through experiments of this kind, the speed of the ions in a given field, and on the other hand—as we can now estimate their electric charge—the force which moves them. They evidently progress more slowly the larger they are; and in the viscous medium constituted by the gas, the displacement is effected at a speed sensibly proportional to the motive power.

At the ordinary temperature these masses are relatively considerable, and are greater for the positive than for the negative ions, that is to say, they are about the order of some ten molecules. The ions, therefore, seem to be formed by an agglomeration of neutral molecules maintained round an electrified centre by electrostatic attraction. If the temperature rises, the thermal agitation will become great enough to prevent the molecules from remaining linked to the centre. By measurements effected on the gases of flames, we arrive at very different values of the masses from those found for ordinary ions, and above all, very different ones for ions of contrary sign. The negative ions have much more considerable velocities than the positive ones. The latter also seem to be of the same size as atoms; and the first-named must, consequently, be considered as very much smaller, and probably about a thousand times less.

Thus, for the first time in science, the idea appears that the atom is not the smallest fraction of matter to be considered. Fragments a thousand times smaller may exist which possess, however, a negative charge. These are the electrons, which other considerations will again bring to our notice.

§ 3. HOW IONS ARE PRODUCED

It is very seldom that a gaseous mass does not contain a few ions. They may have been formed from many causes, for although to give precision to our studies, and to deal with a well ascertained case, I mentioned only ionisation by the X rays in the first instance, I ought not to give the impression that the phenomenon is confined to these rays. It is, on the contrary, very general, and ionisation is just as well produced by the cathode rays, by the radiations emitted by radio-active bodies, by the ultra-violet rays, by heating to a high temperature, by certain chemical actions, and finally by the impact of the ions already existing in neutral molecules.

Of late years these new questions have been the object of a multitude of researches, and if it has not always been possible to avoid some confusion, yet certain general conclusions may be drawn. The ionisation by flames, in particular, is fairly well known. For it to be produced spontaneously, it would appear that there must exist simultaneously a rather high temperature and a chemical action in the gas. According to M. Moreau, the ionisation is very marked when the flame contains the vapour of the salt of an alkali or of an alkaline earth, but much less so when it contains that of other salts.

Arrhenius, Mr C.T.R. Wilson, and M. Moreau, have studied all the circumstances of the phenomenon; and it seems indeed that there is a somewhat close analogy between what first occurs in the saline vapours and that which is noted in liquid electrolytes. There should be produced, as soon as a certain temperature is reached, a dissociation of the saline molecule; and, as M. Moreau has shown in a series of very well conducted researches, the ions formed at about 100°C. seem constituted by an electrified centre of the size of a gas molecule, surrounded by some ten layers of other molecules. We are thus dealing with rather large ions, but according to Mr Wilson, this condensation phenomenon does not affect the number of ions produced by dissociation. In proportion as the temperature rises, the molecules condensed round the nucleus disappear, and, as in all other circumstances, the negative ion tends to become an electron, while the positive ion continues the size of an atom.

In other cases, ions are found still larger than those of saline vapours, as, for example, those produced by phosphorus. It has long been known that air in the neighbourhood of phosphorus becomes a conductor, and the fact, pointed out as far back as 1885 by Matteucci, has been well studied by various experimenters, by MM. Elster and Geitel in 1890, for instance. On the other hand, in 1893 Mr Barus established that the approach of a stick of phosphorus brings about the condensation of water vapour, and we really have before us, therefore, in this instance, an ionisation. M. Bloch has succeeded in disentangling the phenomena, which are here very complex, and in showing that the ions produced are of considerable dimensions; for their speed in the same conditions is on the average a thousand times less than that of ions due to the X rays. M. Bloch has established also that the conductivity of recently-prepared gases, already studied by several authors, was analogous to that which is produced by phosphorus, and that it is intimately connected with the presence of the very tenuous solid or liquid dust which these gases carry with them, while the ions are of the same order of magnitude. These large ions exist, moreover, in small quantities in the atmosphere; and M. Langevin lately succeeded in revealing their presence.

It may happen, and this not without singularly complicating matters, that the ions which were in the midst of material molecules produce, as the result of collisions, new divisions in these last. Other ions are thus born, and this production is in part compensated for by recombinations between ions of opposite signs. The impacts will be more active in the event of the gas being placed in a field of force and of the pressure being slight, the speed attained being then greater and allowing the active force to reach a high value. The energy necessary for the production of an ion is, in fact, according to Professor Rutherford and Professor Stark, something considerable, and it much exceeds the analogous force in electrolytic decomposition.

It is therefore in tubes of rarefied gas that this ionisation by impact will be particularly felt. This gives us the reason for the aspect presented by Geissler tubes. Generally, in the case of discharges, new ions produced by the molecules struck come to add themselves to the electrons produced, as will be seen, by the cathode. A full discussion has led to the interpretation of all the known facts, and to our understanding, for instance, why there exist bright or dark spaces in certain regions of the tube. M. Pellat, in particular, has given some very fine examples of this concordance between the theory and the facts he has skilfully observed.

In all the circumstances, then, in which ions appear, their formation has doubtless been provoked by a mechanism analogous to that of the shock. The X rays, if they are attributable to sudden variations in the ether—that is to say, a variation of the two vectors of Hertz—themselves produce within the atom a kind of electric impulse which breaks it into two electrified fragments; *i.e.* the positive centre, the size of the molecule itself, and the negative centre, constituted by an electron a thousand times smaller. Round these two centres, at the ordinary temperature, are agglomerated by attraction other molecules, and in this manner the ions whose properties have just been studied are formed.

§ 4. ELECTRONS IN METALS

The success of the ionic hypothesis as an interpretation of the conductivity of electrolytes and gases has suggested the desire to try if a similar hypothesis can represent the ordinary conductivity of metals. We are thus led to conceptions which at first sight seem audacious because they are contrary to our habits of mind. They must not, however, be rejected on that account. Electrolytic dissociation at first certainly appeared at least as strange; yet it has ended by forcing itself upon us, and we could, at the present day, hardly dispense with the image it presents to us.

The idea that the conductivity of metals is not essentially different from that of electrolytic liquids or gases, in the sense that the passage of the current is connected with the transport of small electrified particles, is already of old date. It was enunciated by W. Weber, and afterwards developed by Giese, but has only obtained its true scope through the effect of recent discoveries. It was the researches of Riecke, later, of Drude, and, above all, those of J.J. Thomson, which have allowed it to assume an acceptable form. All these attempts are connected however with the general theory of Lorentz, which we will examine later.

It will be admitted that metallic atoms can, like the saline molecule in a solution, partially dissociate themselves. Electrons, very much smaller than atoms, can move through the structure, considerable to them, which is constituted by the atom from which they have just been detached. They may be compared to the molecules of a gas which is enclosed in a porous body. In ordinary conditions, notwithstanding the great speed with which they are animated, they are unable to travel long distances, because they quickly find their road barred by a material atom. They have to undergo innumerable impacts, which throw them first in one direction and then in another. The passage of a current is a sort of flow of these electrons in a determined direction. This electric flow brings, however, no modification to the material medium traversed, since every electron which disappears at any point is replaced by another which appears at once, and in all metals the electrons are identical.

This hypothesis leads us to anticipate certain facts which experience confirms. Thus J.J. Thomson shows that if, in certain conditions, a conductor is placed in a magnetic field, the ions have to describe an epicycloid, and their journey is thus lengthened, while the electric resistance must increase. If the field is in the direction of the displacement, they describe helices round the lines of force and the resistance is again augmented, but in different proportions. Various experimenters have noted phenomena of this kind in different substances.

For a long time it has been noticed that a relation exists between the calorific and the electric conductivity; the relation of these two conductivities is sensibly the same for all metals. The modern theory tends to show simply that it must indeed be so. Calorific conductivity is due, in fact, to an exchange of electrons between the hot and the cold regions, the heated electrons having the greater velocity, and consequently the more considerable energy. The calorific exchanges then obey laws similar to those which govern electric exchanges; and calculation even leads to the exact values which the measurements have given. [31]

In the same way Professor Hesehus has explained how contact electrification is produced, by the tendency of bodies to equalise their superficial properties by means of a transport of electrons, and Mr Jeans has shown that we should discover the existence of the well-known laws of distribution over conducting bodies in electrostatic equilibrium. A metal can, in fact, be electrified, that is to say, may possess an excess of positive or negative electrons which cannot easily leave it in ordinary conditions. To cause them to do so would need an appreciable amount of work, on account of the enormous difference of the specific inductive capacities of the metal and of the insulating medium in which it is plunged.

Electrons, however, which, on arriving at the surface of the metal, possessed a kinetic energy superior to this work, might be shot forth and would be disengaged as a vapour escapes from a liquid. Now, the number of these rapid electrons, at first very slight, increases, according to the kinetic theory, when the temperature rises, and therefore we must reckon that a wire, on being heated, gives out electrons, that is to say, loses negative electricity and sends into the surrounding media electrified centres capable of producing the phenomena of ionisation. Edison, in 1884, showed that from the filament of an incandescent lamp there escaped negative electric charges. Since then, Richardson and J.J. Thomson have examined analogous phenomena. This emission is a very general phenomenon which, no doubt, plays a considerable part in cosmic physics. Professor Arrhenius explains, for instance, the polar auroras by the action of similar corpuscules emitted by the sun.

In other phenomena we seem indeed to be confronted by an emission, not of negative electrons, but of positive ions. Thus, when a wire is heated, not *in vacuo*, but in a gas, this wire begins to electrify neighbouring bodies positively. J.J. Thomson has measured the mass of these positive ions and finds it considerable, *i.e.* about 150 times that of an atom of hydrogen. Some are even larger, and constitute almost a real grain of dust. We here doubtless meet with the phenomena of disaggregation undergone by metals at a red heat.

CHAPTER IX
CATHODE RAYS AND RADIOACTIVE BODIES
§ 1. THE CATHODE RAYS

A wire traversed by an electric current is, as has just been explained, the seat of a movement of electrons. If we cut this wire, a flood of electrons, like a current of water which, at the point where a pipe bursts, flows out in abundance, will appear to spring out between the two ends of the break.

If the energy of the electrons is sufficient, these electrons will in fact rush forth and be propagated in the air or in the insulating medium interposed; but the phenomena of the discharge will in general be very complex. We shall here only examine a particularly simple case, viz., that of the cathode rays; and without entering into details, we shall only note the results relating to these rays which furnish valuable arguments in favour of the electronic hypothesis and supply solid materials for the construction of new theories of electricity and matter.

For a long time it was noticed that the phenomena in a Geissler tube changed their aspect considerably, when the gas pressure became very weak, without, however, a complete vacuum being formed. From the cathode there is shot forth normally and in a straight line a flood within the tube, dark but capable of impressing a photographic plate, of developing the fluorescence of various substances (particularly the glass walls of the tube), and of producing calorific and mechanical effects. These are the cathode rays, so named in 1883 by E. Wiedemann, and their name, which was unknown to a great number of physicists till barely twelve years ago, has become popular at the present day.

About 1869, Hittorf made an already very complete study of them and put in evidence their principal properties; but it was the researches of Sir W. Crookes in especial which drew attention to them. The celebrated physicist foresaw that the phenomena which were thus produced in rarefied gases were, in spite of their very great complication, more simple than those presented by matter under the conditions in which it is generally met with.

He devised a celebrated theory no longer admissible in its entirety, because it is not in complete accord with the facts, which was, however, very interesting, and contained, in germ, certain of our present ideas. In the opinion of Crookes, in a tube in which the gas has been rarefied we are in presence of a special state of matter. The number of the gas molecules has become small enough for their independence to be almost absolute, and they are able in this so-called radiant state to traverse long spaces without departing from a straight line. The cathode rays are due to a kind of molecular bombardment of the walls of the tubes, and of the screens which can be introduced into them; and it is the molecules, electrified by their contact with the cathode and then forcibly repelled by electrostatic action, which produce, by their movement and their *vis viva*, all the phenomena observed. Moreover, these electrified molecules animated with extremely rapid velocities correspond, according to the theory verified in the celebrated experiment of Rowland on convection currents, to a true electric current, and can be deviated by a magnet.

Notwithstanding the success of Crookes' experiments, many physicists—the Germans especially—did not abandon an hypothesis entirely different from that of radiant matter. They continued to regard the cathode radiation as due to particular radiations of a nature still little known but produced in the luminous ether. This interpretation seemed, indeed, in 1894, destined to triumph definitely through the remarkable discovery of Lenard, a discovery which, in its turn, was to provoke so many others and to bring about consequences of which the importance seems every day more considerable.

Professor Lenard's fundamental idea was to study the cathode rays under conditions different from those in which they are produced. These rays are born in a very rarefied space, under conditions perfectly determined by Sir W. Crookes; but it was a question whether, when once produced, they would be capable of propagating themselves in other media, such as a gas at ordinary pressure, or even in an absolute vacuum. Experiment alone could answer this question, but there were difficulties in the way of this which seemed almost insurmountable. The rays are stopped by glass even of slight thickness, and how then could the almost vacuous space in which they have to come into existence be separated from the space, absolutely vacuous or filled with gas, into which it was desired to bring them?

The artifice used was suggested to Professor Lenard by an experiment of Hertz. The great physicist had, in fact, shortly before his premature death, taken up this important question of the cathode rays, and his genius left there, as elsewhere, its powerful impress. He had shown that metallic plates of very slight thickness were transparent to the cathode rays; and Professor Lenard succeeded in obtaining plates impermeable to air, but which yet allowed the pencil of cathode rays to pass through them.

Now if we take a Crookes tube with the extremity hermetically closed by a metallic plate with a slit across the diameter of 1 mm. in width, and stop this slit with a sheet of very thin aluminium, it will be immediately noticed that the rays pass through the aluminium and pass outside the tube. They are propagated in air at atmospheric pressure, and they can also penetrate into an absolute vacuum. They therefore can no longer be attributed to radiant matter, and we are led to think that the energy brought into play in this phenomenon must have its seat in the light-bearing ether itself.

But it is a very strange light which is thus subject to magnetic action, which does not obey the principle of equal angles, and for which the most various gases are already disturbed media. According to Crookes it possesses also the singular property of carrying with it electric charges.

This convection of negative electricity by the cathode rays seems quite inexplicable on the hypothesis that the rays are ethereal radiations. Nothing then remained in order to maintain this hypothesis, except to deny the convection, which, besides, was only established by indirect experiments. That the reality of this transport has been placed beyond dispute by means of an extremely elegant experiment which is all the more convincing that it is so very simple, is due to M. Perrin. In the interior of a Crookes tube he collected a pencil of cathode rays in a metal cylinder. According to the elementary principles of electricity the cylinder must become charged with the whole charge, if there be one, brought to it by the rays, and naturally various precautions had to be taken. But the result was very precise, and doubt could no longer exist—the rays were electrified.

It might have been, and indeed was, maintained, some time after this experiment was published, that while the phenomena were complex inside the tube, outside, things might perhaps occur differently. Lenard himself, however, with that absence of even involuntary prejudice common to all great minds, undertook to demonstrate that the opinion he at first held could no longer be accepted, and succeeded in repeating the experiment of M. Perrin on cathode rays in the air and even *in vacuo*.

On the wrecks of the two contradictory hypotheses thus destroyed, and out of the materials from which they had been built, a theory has been constructed which co-ordinates all the known facts. This theory is furthermore closely allied to the theory of ionisation, and, like this latter, is based on the concept of the electron. Cathode rays are electrons in rapid motion.

The phenomena produced both inside and outside a Crookes tube are, however, generally complex. In Lenard's first experiments, and in many others effected later when this region of physics was still very little known, a few confusions may be noticed even at the present day.

At the spot where the cathode rays strike the walls of the tube the essentially different X rays appear. These differ from the cathode radiations by being neither electrified nor deviated by a magnet. In their turn these X rays may give birth to the secondary rays of M. Sagnac; and often we find ourselves in presence of effects from these last-named radiations and not from the true cathode rays.

The electrons, when they are propagated in a gas, can ionise the molecules of this gas and unite with the neutral atoms to form negative ions, while positive ions also appear. There are likewise produced, at the expense of the gas still subsisting after rarefication within the tube, positive ions which, attracted by the cathode and reaching it, are not all neutralised by the negative electrons, and can, if the cathode be perforated, pass through it, and if not, pass round it. We have then what are called the canal rays of Goldstein, which are deviated by an electric or magnetic field in a contrary direction to the cathode rays; but, being larger, give weak deviations or may even remain undeviated through losing their charge when passing through the cathode.

It may also be the parts of the walls at a distance from the cathode which send a positive rush to the latter, by a similar mechanism. It may be, again, that in certain regions of the tube cathode rays are met with diffused by some solid object, without having thereby changed their nature. All these complexities have been cleared up by M. Villard, who has published, on these questions, some remarkably ingenious and particularly careful experiments.

M. Villard has also studied the phenomena of the coiling of the rays in a field, as already pointed out by Hittorf and Plücker. When a magnetic field acts on the cathode particle, the latter follows a trajectory, generally helicoidal, which is anticipated by the theory. We here have to do with a question of ballistics, and experiments duly confirm the anticipations of the calculation. Nevertheless, rather singular phenomena appear in the case of certain values of the field, and these phenomena, dimly seen by Plücker and Birkeland, have been the object of experiments by M. Villard. The two faces of the cathode seem to emit rays which are deviated in a direction perpendicular to the lines of force by an electric field, and do not seem to be electrified. M. Villard calls them magneto-cathode rays, and according to M. Fortin these rays may be ordinary cathode rays, but of very slight velocity.

In certain cases the cathode itself may be superficially disaggregated, and extremely tenuous particles detach themselves, which, being carried off at right angles to its surface, may deposit themselves like a very thin film on objects placed in their path. Various physicists, among them M. Houllevigue, have studied this phenomenon, and in the case of pressures between 1/20 and 1/100 of a millimetre, the last-named scholar has obtained mirrors of most metals, a phenomenon he designates by the name of ionoplasty.

But in spite of all these accessory phenomena, which even sometimes conceal those first observed, the existence of the electron in the cathodic flux remains the essential characteristic.

The electron can be apprehended in the cathodic ray by the study of its essential properties; and J.J. Thomson gave great value to the hypothesis by his measurements. At first he meant to determine the speed of the cathode rays by direct experiment, and by observing, in a revolving mirror, the relative displacement of two bands due to the excitement of two fluorescent screens placed at different distances from the cathode. But he soon perceived that the effect of the fluorescence was not instantaneous, and that the lapse of time might form a great source of error, and he then had recourse to indirect methods. It is possible, by a simple calculation, to estimate the deviations produced on the rays by a magnetic and an electric field respectively as a function of the speed of propagation and of the relation of the charge to the material mass of the electron. The measurement of these deviations will then permit this speed and this relation to be ascertained.

Other processes may be used which all give the same two quantities by two suitably chosen measurements. Such are the radius of the curve taken by the trajectory of the pencil in a perpendicular magnetic field and the measure of the fall of potential under which the discharge takes place, or the measure of the total quantity of electricity carried in one second and the measure of the calorific energy which may be given, during the same period, to a thermo-electric junction. The results agree as well as can be expected, having regard to the difficulty of the experiments; the values of the speed agree also with those which Professor Wiechert has obtained by direct measurement.

The speed never depends on the nature of the gas contained in the Crookes tube, but varies with the value of the fall of potential at the cathode. It is of the order of one tenth of the speed of light, and it may rise as high as one third. The cathode particle therefore goes about three thousand times faster than the earth in its orbit. The relation is also invariable, even when the substance of which the cathode is formed is changed or one gas is substituted for another. It is, on the average, a thousand times greater than the corresponding relation in electrolysis. As experiment has shown, in all the circumstances where it has been possible to effect measurements, the equality of the charges carried by all corpuscules, ions, atoms, etc., we ought to consider that the charge of the electron is here, again, that of a univalent ion in electrolysis, and therefore that its mass is only a small fraction of that of the atom of hydrogen, viz., of the order of about a thousandth part. This is the same result as that to which we were led by the study of flames.

The thorough examination of the cathode radiation, then, confirms us in the idea that every material atom can be dissociated and will yield an electron much smaller than itself—and always identical whatever the matter whence it comes,—the rest of the atom remaining charged with a positive quantity equal and contrary to that borne by the electron. In the present case these positive ions are no doubt those that we again meet with in the canal rays. Professor Wien has shown that their mass is really, in fact, of the order of the mass of atoms. Although they are all formed of identical electrons, there may be various cathode rays, because the velocity is not exactly the same for all electrons. Thus is explained the fact that we can separate them and that we can produce a sort of spectrum by the action of the magnet, or, again, as M. Deslandres has shown in a very interesting experiment, by that of an electrostatic field. This also probably explains the phenomena studied by M. Villard, and previously pointed out.

§ 2. RADIOACTIVE SUBSTANCES

Even in ordinary conditions, certain substances called radioactive emit, quite outside any particular reaction, radiations complex indeed, but which pass through fairly thin layers of minerals, impress photographic plates, excite fluorescence, and ionize gases. In these radiations we again find electrons which thus escape spontaneously from radioactive bodies.

It is not necessary to give here a history of the discovery of radium, for every one knows the admirable researches of M. and Madame Curie. But subsequent to these first studies, a great number of facts have accumulated for the last six years, among which some people find themselves a little lost. It may, perhaps, not be useless to indicate the essential results actually obtained.

The researches on radioactive substances have their starting-point in the discovery of the rays of uranium made by M. Becquerel in 1896. As early as 1867 Niepce de St Victor proved that salts of uranium impressed photographic plates in the dark; but at that time the phenomenon could only pass for a singularity attributable to phosphorescence, and the valuable remarks of Niepce fell into oblivion. M. Becquerel established, after some hesitations natural in the face of phenomena which seemed so contrary to accepted ideas, that the radiating property was absolutely independent of phosphorescence, that all the salts of uranium, even the uranous salts which are not phosphorescent, give similar radiant effects, and that these phenomena correspond to a continuous emission of energy, but do not seem to be the result of a storage of energy under the influence of some external radiation. Spontaneous and constant, the radiation is insensible to variations of temperature and light.

The nature of these radiations was not immediately understood, [32] and their properties seemed contradictory. This was because we were not dealing with a single category of rays. But amongst all the effects there is one which constitutes for the radiations taken as a whole, a veritable process for the measurement of radioactivity. This is their ionizing action on gases. A very complete study of the conductivity of air under the influence of rays of uranium has been made by various physicists, particularly by Professor Rutherford, and has shown that the laws of the phenomenon are the same as those of the ionization due to the action of the Röntgen rays.

It was natural to ask one's self if the property discovered in salts of uranium was peculiar to this body, or if it were not, to a more or less degree, a general property of matter. Madame Curie and M. Schmidt, independently of each other, made systematic researches in order to solve the question; various compounds of nearly all the simple bodies at present known were thus passed in review, and it was established that radioactivity was particularly perceptible in the compounds of uranium and thorium, and that it was an atomic property linked to the matter endowed with it, and following it in all its combinations. In the course of her researches Madame Curie observed that certain pitchblendes (oxide of uranium ore, containing also barium, bismuth, etc.) were four times more active (activity being measured by the phenomenon of the ionization of the air) than metallic uranium. Now, no compound containing any other active metal than uranium or thorium ought to show itself more active than those metals themselves, since the property belongs to their atoms. It seemed, therefore, probable that there existed in pitchblendes some substance yet unknown, in small quantities and more radioactive than uranium.

M. and Madame Curie then commenced those celebrated experiments which brought them to the discovery of radium. Their method of research has been justly compared in originality and importance to the process of spectrum analysis. To isolate a radioactive substance, the first thing is to

173

measure the activity of a certain compound suspected of containing this substance, and this compound is chemically separated. We then again take in hand all the products obtained, and by measuring their activity anew, it is ascertained whether the substance sought for has remained in one of these products, or is divided among them, and if so, in what proportion. The spectroscopic reaction which we may use in the course of this separation is a thousand times less sensitive than observation of the activity by means of the electrometer.

Though the principle on which the operation of the concentration of the radium rests is admirable in its simplicity, its application is nevertheless very laborious. Tons of uranium residues have to be treated in order to obtain a few decigrammes of pure salts of radium. Radium is characterised by a special spectrum, and its atomic weight, as determined by Madame Curie, is 225; it is consequently the higher homologue of barium in one of the groups of Mendeléef. Salts of radium have in general the same chemical properties as the corresponding salts of barium, but are distinguished from them by the differences of solubility which allow of their separation, and by their enormous activity, which is about a hundred thousand times greater than that of uranium.

Radium produces various chemical and some very intense physiological reactions. Its salts are luminous in the dark, but this luminosity, at first very bright, gradually diminishes as the salts get older. We have here to do with a secondary reaction correlative to the production of the emanation, after which radium undergoes the transformations which will be studied later on.

The method of analysis founded by M. and Madame Curie has enabled other bodies presenting sensible radioactivity to be discovered. The alkaline metals appear to possess this property in a slight degree. Recently fallen snow and mineral waters manifest marked action. The phenomenon may often be due, however, to a radioactivity induced by radiations already existing in the atmosphere. But this radioactivity hardly attains the ten-thousandth part of that presented by uranium, or the ten-millionth of that appertaining to radium.

Two other bodies, polonium and actinium, the one characterised by the special nature of the radiations it emits and the other by a particular spectrum, seem likewise to exist in pitchblende. These chemical properties have not yet been perfectly defined; thus M. Debierne, who discovered actinium, has been able to note the active property which seems to belong to it, sometimes in lanthanum, sometimes in neodynium.[33] It is proved that all extremely radioactive bodies are the seat of incessant transformations, and even now we cannot state the conditions under which they present themselves in a strictly determined form.

§ 3. THE RADIATION OF THE RADIOACTIVE BODIES AND THE EMANATION

To acquire exact notions as to the nature of the rays emitted by the radioactive bodies, it was necessary to try to cause magnetic or electric forces to act on them so as to see whether they behaved in the same way as light and the X rays, or whether like the cathode rays they were deviated by a magnetic field. This work was effected by Professor Giesel, then by M. Becquerel, Professor Rutherford, and by many other experimenters after them. All the methods which have already been mentioned in principle have been employed in order to discover whether they were electrified, and, if so, by electricity of what sign, to measure their speed, and to ascertain their degree of penetration.

The general result has been to distinguish three sorts of radiations, designated by the letters alpha, beta, gamma.

The alpha rays are positively charged, and are projected at a speed which may attain the tenth of that of light; M.H. Becquerel has shown by the aid of photography that they are deviated by a magnet, and Professor Rutherford has, on his side, studied this deviation by the electrical method. The relation of the charge to the mass is, in the case of these rays, of the same order as in that of the ions of electrolysis. They may therefore be considered as exactly analogous to the canal rays of Goldstein, and we may attribute them to a material transport of corpuscles of the magnitude of atoms. The relatively considerable size of these corpuscles renders them very absorbable. A flight of a few millimetres in a gas suffices to reduce their number by one-half. They have great ionizing power.

The beta rays are on all points similar to the cathode rays; they are, as M. and Madame Curie have shown, negatively charged, and the charge they carry is always the same. Their size is that of the electrons, and their velocity is generally greater than that of the cathode rays, while it may become almost that of light. They have about a hundred times less ionizing power than the alpha rays.

The gamma rays were discovered by M. Villard.[34] They may be compared to the X rays; like the latter, they are not deviated by the magnetic field, and are also extremely penetrating. A strip of aluminium five millimetres thick will stop the other kinds, but will allow them to pass. On the other hand, their ionizing power is 10,000 times less than that of the alpha rays.

To these radiations there sometimes are added in the course of experiments secondary radiations analogous to those of M. Sagnac, and produced when the alpha, beta, or gamma rays meet various substances. This complication has often led to some errors of observation.

Phosphorescence and fluorescence seem especially to result from the alpha and beta rays, particularly from the alpha rays, to which belongs the most important part of the total energy of the radiation. Sir W. Crookes has invented a curious little apparatus, the spinthariscope, which enables us to examine the phosphorescence of the blende excited by these rays. By means of a magnifying glass, a screen covered with sulphide of zinc is kept under observation, and in front of it is disposed, at a distance of about half a millimetre, a fragment of some salt of radium. We then perceive multitudes of brilliant points on the screen, which appear and at once disappear, producing a scintillating effect. It seems probable that every particle falling on the screen produces by its impact a disturbance in the neighbouring region, and it is this disturbance which the eye perceives as a luminous point. Thus, says Sir W. Crookes, each drop of rain falling on the surface of still water is not perceived as a drop of rain, but by reason of the slight splash which it causes at the moment of impact, and which is manifested by ridges and waves spreading themselves in circles.

The various radioactive substances do not all give radiations of identical constitution. Radium and thorium possess in somewhat large proportions the three kinds of rays, and it is the same with actinium. Polonium contains especially alpha rays and a few gamma rays. [35] In the case of uranium, the alpha rays have extremely slight penetrating power, and cannot even impress photographic plates. But the widest difference between the substances proceeds from the emanation. Radium, in addition to the three groups of rays alpha, beta, and gamma, disengages continuously an extremely subtle emanation, seemingly almost imponderable, but which may be, for many reasons, looked upon as a vapour of which the elastic force is extremely feeble.

M. and Madame Curie discovered as early as 1899 that every substance placed in the neighbourhood of radium, itself acquired a radioactivity which persisted for several hours after the removal of the radium. This induced radioactivity seems to be carried to other bodies by the intermediary of a gas. It goes round obstacles, but there must exist between the radium and the substance a free and continuous space for the activation to take place; it cannot, for instance, do so through a wall of glass.

In the case of compounds of thorium Professor Rutherford discovered a similar phenomenon; since then, various physicists, Professor Soddy, Miss Brooks, Miss Gates, M. Danne, and others, have studied the properties of these emanations.

The substance emanated can neither be weighed nor can its elastic force be ascertained; but its transformations may be followed, as it is luminous, and it is even more certainly characterised by its essential property, *i.e.* its radioactivity. We also see that it can be decanted like a gas, that it will divide itself between two tubes of different capacity in obedience to the law of Mariotte, and will condense in a refrigerated tube in accordance with the principle of Watt, while it even complies with the law of Gay-Lussac.

The activity of the emanation vanishes quickly, and at the end of four days it has diminished by one-half. If a salt of radium is heated, the emanation becomes more abundant, and the residue, which, however, does not sensibly diminish in weight, will have lost all its radioactivity, and will only recover it by degrees. Professor Rutherford, notwithstanding many different attempts, has been unable to make this emanation enter into any chemical reaction. If it be a gaseous body, it must form part of the argon group, and, like its other members, be perfectly inert.

By studying the spectrum of the gas disengaged by a solution of salt of radium, Sir William Ramsay and Professor Soddy remarked that when the gas is radioactive there are first obtained rays of gases belonging to the argon family, then by degrees, as the activity disappears, the spectrum slowly changes, and finally presents the characteristic aspect of helium.

We know that the existence of this gas was first discovered by spectrum analysis in the sun. Later its presence was noted in our atmosphere, and in a few minerals which happen to be the very ones from which radium has been obtained. It might therefore have been the case that it pre-existed in the gases extracted from radium; but a remarkable experiment by M. Curie and Sir James Dewar seems to show convincingly that this cannot be so. The spectrum of helium never appears at first in the gas proceeding from pure bromide of radium; but it shows itself, on the other hand, very distinctly, after the radioactive transformations undergone by the salt.

All these strange phenomena suggest bold hypotheses, but to construct them with any solidity they must be supported by the greatest possible number of facts. Before admitting a definite explanation of the phenomena which have their seat in the curious substances discovered by them, M. and Madame Curie considered, with a great deal of reason, that they ought first to enrich our knowledge with the exact and precise facts relating to these bodies and to the effects produced by the radiations they emit.

Thus M. Curie particularly set himself to study the manner in which the radioactivity of the emanation is dissipated, and the radioactivity that this emanation can induce on all bodies. The radioactivity of the emanation diminishes in accordance with an exponential law. The constant of time which characterises this decrease is easily and exactly determined, and has a fixed value, independent of the conditions of the experiment as well as of the nature of the gas which is in contact with the radium and becomes charged with the emanation. The regularity of the phenomenon is so great that it can be used to measure time: in 3985 seconds [36] the activity is always reduced one-half.

Radioactivity induced on any body which has been for a long time in presence of a salt of radium disappears more rapidly. The phenomenon appears, moreover, more complex, and the formula which expresses the manner in which the activity diminishes must contain two exponentials. To find it theoretically we have to imagine that the emanation first deposits on the body in question a substance which is destroyed in giving birth to a second, this latter disappearing in its turn by generating a third. The initial and final substances would be radioactive, but the intermediary one, not. If, moreover, the bodies acted on are brought to a temperature of over 700°, they appear to lose by volatilisation certain substances condensed in them, and at the same time their activity disappears.

The other radioactive bodies behave in a similar way. Bodies which contain actinium are particularly rich in emanations. Uranium, on the contrary, has none.[37] This body, nevertheless, is the seat of transformations comparable to those which the study of emanations reveals in radium; Sir W. Crookes has separated from uranium a matter which is now called uranium X. This matter is at first much more active than its parent, but its activity diminishes rapidly, while the ordinary uranium, which at the time of the separation loses its activity, regains it by degrees. In the same way, Professors Rutherford and Soddy have discovered a so-called thorium X to be the stage through which ordinary thorium has to pass in order to produce its emanation. [38]

It is not possible to give a complete table which should, as it were, represent the genealogical tree of the various radioactive substances. Several authors have endeavoured to do so, but in a premature manner; all the affiliations are not at the present time yet perfectly known, and it will no doubt be acknowledged some day that identical states have been described under different names. [39]

§ 4. THE DISAGGREGATION OF MATTER AND ATOMIC ENERGY

In spite of uncertainties which are not yet entirely removed, it cannot be denied that many experiments render it probable that in radioactive bodies we find ourselves witnessing veritable transformations of matter.

Professor Rutherford, Professor Soddy, and several other physicists, have come to regard these phenomena in the following way. A radioactive body is composed of atoms which have little stability, and are able to detach themselves spontaneously from the parent substance, and at the same time to divide themselves into two essential component parts, the negative electron and its residue the positive ion. The first-named constitutes the beta, and the second the alpha rays.

The emanation is certainly composed of alpha ions with a few molecules agglomerated round them. Professor Rutherford has, in fact, demonstrated that the emanation is charged with positive electricity; and this emanation may, in turn, be destroyed by giving birth to new bodies.

After the loss of the atoms which are carried off by the radiation, the remainder of the body acquires new properties, but it may still be radioactive, and again lose atoms. The various stages that we meet with in the evolution of the radioactive substance or of its emanation, correspond to the various degrees of atomic disaggregation. Professors Rutherford and Soddy have described them clearly in the case of uranium and radium. As regards thorium the results are less satisfactory. The evolution should continue until a stable atomic condition is finally reached, which, because of this stability, is no longer radioactive. Thus, for instance, radium would finally be transformed into helium.[40]

It is possible, by considerations analogous to those set forth above in other cases, to arrive at an idea of the total number of particles per second expelled by one gramme of radium; Professor Rutherford in his most recent evaluation finds that this number approaches 2.5×10^{11}.[41] By calculating from the atomic weight the number of atoms probably contained in this gramme of radium, and supposing each particle liberated to correspond to the destruction of one atom, it is found that one half of the radium should disappear in 1280 years;[42] and from this we may conceive that it has not yet been possible to discover any sensible loss of weight. Sir W. Ramsay and Professor Soddy attained a like result by endeavouring to estimate the mass of the emanation by the quantity of helium produced.

If radium transforms itself in such a way that its activity does not persist throughout the ages, it loses little by little the provision of energy it had in the beginning, and its properties furnish no valid argument to oppose to the principle of the conservation of energy. To put everything right, we have only to recognise that radium possessed in the potential state at its formation a finite quantity of energy which is consumed little by little. In the same manner, a chemical system composed, for instance, of zinc and sulphuric acid, also contains in the potential state energy which, if we retard the reaction by any suitable arrangement—such as by amalgamating the zinc and by constituting with its elements a battery which we cause to act on a resistance—may be made to exhaust itself as slowly as one may desire.

There can, therefore, be nothing in any way surprising in the fact that a combination which, like the atomic combination of radium, is not stable—since it disaggregates itself,—is capable of spontaneously liberating energy, but what may be a little astonishing, at first sight, is the considerable amount of this energy.

M. Curie has calculated directly, by the aid of the calorimeter, the quantity of energy liberated, measuring it entirely in the form of heat. The disengagement of heat accounted for in a grain of radium is uniform, and amounts to 100 calories per hour. It must therefore be admitted that an atom of radium, in disaggregating itself, liberates 30,000 times more energy than a molecule of hydrogen when the latter combines with an atom of oxygen to form a molecule of water.

We may ask ourselves how the atomic edifice of the active body can be constructed, to contain so great a provision of energy. We will remark that such a question might be asked concerning cases known from the most remote antiquity, like that of the chemical systems, without any satisfactory answer ever being given. This failure surprises no one, for we get used to everything—even to defeat.

When we come to deal with a new problem we have really no right to show ourselves more exacting; yet there are found persons who refuse to admit the hypothesis of the atomic disaggregation of radium because they cannot have set before them a detailed plan of that complex whole known to us as an atom.

The most natural idea is perhaps the one suggested by comparison with those astronomical phenomena where our observation most readily allows us to comprehend the laws of motion. It corresponds likewise to the tendency ever present in the mind of man, to compare the infinitely small with the infinitely great. The atom may be regarded as a sort of solar system in which electrons in considerable numbers gravitate round the sun formed by the positive ion. It may happen that certain of these electrons are no longer retained in their orbit by the electric attraction of the rest of the atom, and may be projected from it like a small planet or comet which escapes towards the stellar spaces. The phenomena of the emission of light compels us to think that the corpuscles revolve round the nucleus with extreme velocities, or at the rate of thousands of billions of evolutions per second. It is easy to conceive from this that, notwithstanding its lightness, an atom thus constituted may possess an enormous energy.[43]

Other authors imagine that the energy of the corpuscles is principally due to the extremely rapid rotations of those elements on their own axes. Lord Kelvin lately drew up on another model the plan of a radioactive atom capable of ejecting an electron with a considerable *vis viva*. He supposes a spherical atom formed of concentric layers of positive and negative electricity disposed in such a way that its external action is null, and that, nevertheless, the force emanated from the centre may be repellent for certain values when the electron is within it.

The most prudent physicists and those most respectful to established principles may, without any scruples, admit the explanation of the radioactivity of radium by a dislocation of its molecular edifice. The matter of which it is constituted evolves from an admittedly unstable initial state to another stable one. It is, in a way, a slow allotropic transformation which takes place by means of a mechanism regarding which, in short, we have no more information than we have regarding other analogous transformations. The only astonishment we can legitimately feel is derived from the thought that we are suddenly and deeply penetrating to the very heart of things.

But those persons who have a little more hardihood do not easily resist the temptation of forming daring generalisations. Thus it will occur to some that this property, already discovered in many substances where it exists in more or less striking degree, is, with differences of intensity, common to all bodies, and that we are thus confronted by a phenomenon derived from an essential quality of matter. Quite recently, Professor Rutherford has demonstrated in a fine series of experiments that the alpha particles of radium cease to ionize gases when they are made to lose their velocity, but that they do not on that account cease to exist. It may follow that many bodies emit similar particles without being easily perceived to do so; since the electric action, by which this phenomenon of radioactivity is generally manifested, would, in this case, be but very weak.

If we thus believe radioactivity to be an absolutely general phenomenon, we find ourselves face to face with a new problem. The transformation of radioactive bodies can no longer be assimilated to allotropic transformations, since thus no final form could ever be attained, and the disaggregation would continue indefinitely up to the complete dislocation of the atom. [44] The phenomenon might, it is true, have a duration of perhaps thousands of millions of centuries, but this duration is but a minute in the infinity of time, and matters little. Our habits of mind, if we adopt such a conception, will be none the less very deeply disturbed. We shall have to abandon the idea so instinctively dear to us that matter is the most stable thing in the universe, and to admit, on the contrary, that all bodies whatever are a kind of explosive decomposing with extreme slowness. There is in this, whatever may have been said, nothing contrary to any of the principles on which the science of energetics rests; but an hypothesis of this nature carries with it consequences which ought in the highest degree to interest the philosopher, and we all know with what alluring boldness M. Gustave Le Bon has developed all these consequences in his work on the evolution of matter.[45]

There is hardly a physicist who does not at the present day adopt in one shape or another the ballistic hypothesis. All new facts are co-ordinated so happily by it, that it more and more satisfies our minds; but it cannot be asserted that it forces itself on our convictions with irresistible weight. Another point of view appeared more plausible and simple at the outset, when there seemed reason to consider the energy radiated by radioactive bodies as inexhaustible. It was thought that the source of this energy was to be looked for without the atom, and this idea may perfectly well he maintained at the present day.

Radium on this hypothesis must be considered as a transformer borrowing energy from the external medium and returning it in the form of radiation. It is not impossible, even, to admit that the energy which the atom of radium withdraws from the surrounding medium may serve to keep up, not only the heat emitted and its complex radiation, but also the dissociation, supposed to be endothermic, of this atom. Such seems to be the idea of M. Debierne and also of M. Sagnac. It does not seem to accord with the experiments that this borrowed energy can be a part of the heat of the ambient medium; and, indeed, such a phenomenon would be contrary to the principle of Carnot if we wished (though we have seen how disputable is this extension) to extend this principle to the phenomena which are produced in the very bosom of the atom.

We may also address ourselves to a more noble form of energy, and ask ourselves whether we are not, for the first time, in presence of a transformation of gravitational energy. It may be singular, but it is not absurd, to suppose that the unit of mass of radium is not attached to the earth with the same intensity as an inert body. M. Sagnac has commenced some experiments, as yet unpublished, in order to study the laws of the fall of a fragment of radium. They are necessarily very delicate, and the energetic and ingenious physicist has not yet succeeded in finishing them. [46] Let us suppose that he succeeds in demonstrating that the intensity of gravity is less for radium than for the platinum or the copper of which the pendulums used to illustrate the law of Newton are generally made; it would then be possible still to think that the laws of universal attraction are perfectly exact as regards the stars, and that ponderability is really a particular case of universal attraction, while in the case of radioactive bodies part of the gravitational energy is transformed in the course of its evolution and appears in the form of active radiation.

But for this explanation to be admitted, it would evidently need to be supported by very numerous facts. It might, no doubt, appear still more probable that the energy borrowed from the external medium by radium is one of those still unknown to us, but of which a vague instinct causes us to suspect the existence around us. It is indisputable, moreover, that the atmosphere in all directions is furrowed with active radiations; those of radium may be secondary radiations reflected by a kind of resonance phenomenon.

Certain experiments by Professors Elster and Geitel, however, are not favourable to this point of view. If an active body be surrounded by a radioactive envelope, a screen should prevent this body from receiving any impression from outside, and yet there is no diminution apparent in the activity presented by a certain quantity of radium when it is lowered to a depth of 800 metres under ground, in a region containing a notable quantity of pitchblende. These negative results are, on the other hand, so many successes for the partisans of the explanation of radioactivity by atomic energy.

CHAPTER X
THE ETHER AND MATTER
§ 1. THE RELATIONS BETWEEN THE ETHER AND MATTER

For some time past it has been the more or less avowed ambition of physicists to construct with the particles of ether all possible forms of corporeal existence; but our knowledge of the inmost nature of things has hitherto seemed too limited for us to attempt such an enterprise with any chance of success. The electronic hypothesis, however, which has furnished a satisfactory image of the most curious phenomena produced in the bosom of matter, has also led to a more complete electromagnetic theory of the ether than that of Maxwell, and this twofold result has given birth to the hope of arriving by means of this hypothesis at a complete co-ordination of the physical world.

The phenomena whose study may bring us to the very threshold of the problem, are those in which the connections between matter and the ether appear clearly and in a relatively simple manner. Thus in the phenomena of emission, ponderable matter is seen to give birth to waves which are transmitted by the ether, and by the phenomena of absorption it is proved that these waves disappear and excite modifications in the interior of the material bodies which receive them. We here catch in operation actual reciprocal actions and reactions between the ether and matter. If we could thoroughly comprehend these actions, we should no doubt be in a position to fill up the gap which separates the two regions separately conquered by physical science.

In recent years numerous researches have supplied valuable materials which ought to be utilized by those endeavouring to construct a theory of radiation. We are, perhaps, still ill informed as to the phenomena of luminescence in which undulations are produced in a complex manner, as in the case of a stick of moist phosphorus which is luminescent in the dark, or in that of a fluorescent screen. But we are very well acquainted with emission or absorption by incandescence, where the only transformation is that of calorific into radiating energy, or *vice versa*. It is in this case alone that can be correctly applied the celebrated demonstration by which Kirchhoff established, by considerations borrowed from thermodynamics, the proportional relations between the power of emission and that of absorption.

In treating of the measurement of temperature, I have already pointed out the experiments of Professors Lummer and Pringsheim and the theoretical researches of Stephan and Professor Wien. We may consider that at the present day the laws of the radiation of dark bodies are tolerably well known, and, in particular, the manner in which each elementary radiation increases with the temperature. A few doubts, however, subsist with respect to the law of the distribution of energy in the spectrum. In the case of real and solid bodies the results are naturally less simple than in that of dark bodies. One side of the question has been specially studied on account of its great practical interest, that is to say, the fact that the relation of the luminous energy to the total amount radiated by a body varies with the nature of this last; and the knowledge of the conditions under which this relation becomes most considerable led to the discovery of incandescent lighting by gas in the Auer-Welsbach mantle, and to the substitution for the carbon thread in the electric light bulb of a filament of osmium or a small rod of magnesium, as in the Nernst lamp. Careful measurements effected by M. Fery have furnished, in particular, important information on the radiation of the white oxides; but the phenomena noticed have not yet found a satisfactory interpretation. Moreover, the radiation of calorific origin is here accompanied by a more or less important luminescence, and the problem becomes very complex.

In the same way that, for the purpose of knowing the constitution of matter, it first occurred to us to investigate gases, which appear to be molecular edifices built on a more simple and uniform plan than solids, we ought naturally to think that an examination of the conditions in which emission and absorption are produced by gaseous bodies might be eminently profitable, and might perhaps reveal the mechanism by which the relations between the molecule of the ether and the molecule of matter might be established.

Unfortunately, if a gas is not absolutely incapable of emitting some sort of rays by simple heat, the radiation thus produced, no doubt by reason of the slightness of the mass in play, always remains of moderate intensity. In nearly all the experiments, new energies of chemical or electrical origin come into force. On incandescence, luminescence is superposed; and the advantage which might have been expected from the simplicity of the medium vanishes through the complication of the circumstances in which the phenomenon is produced.

Professor Pringsheim has succeeded, in certain cases, in finding the dividing line between the phenomena of luminescence and that of incandescence. Thus the former takes a predominating importance when the gas is rendered luminous by electrical discharges, and chemical transformations, especially, play a preponderant rôle in the emission of the spectrum of flames which contain a saline vapour. In all the ordinary experiments of spectrum analysis the laws of Kirchhoff cannot therefore be considered as established, and yet the relation between emission and absorption is generally tolerably well verified. No doubt we are here in presence of a kind of resonance phenomenon, the gaseous atoms entering into vibration when solicited by the ether by a motion identical with the one they are capable of communicating to it.

If we are not yet very far advanced in the study of the mechanism of the production of the spectrum,[47] we are, on the other hand, well acquainted with its constitution. The extreme confusion which the spectra of the lines of the gases seemed to present is now, in great part at least, cleared up. Balmer gave some time since, in the case of the hydrogen spectrum, an empirical formula which enabled the rays discovered later by an eminent astronomer, M. Deslandres, to be represented; but since then, both in the cases of line and band spectra, the labours of Professor Rydberg, of M. Deslandres, of Professors Kayzer and Runge, and of M. Thiele, have enabled us to comprehend, in their smallest details, the laws of the distribution of lines and bands.

These laws are simple, but somewhat singular. The radiations emitted by a gas cannot be compared to the notes to which a sonorous body gives birth, nor even to the most complicated vibrations of any elastic body. The number of vibrations of the different rays are not the successive multiples of one and the same number, and it is not a question of a fundamental radiation and its harmonics, while—and this is an essential difference—the number of vibrations of the radiation tend towards a limit when the period diminishes infinitely instead of constantly increasing, as would be the case with the vibrations of sound.

Thus the assimilation of the luminous to the elastic vibration is not correct. Once again we find that the ether does not behave like matter which obeys the ordinary laws of mechanics, and every theory must take full account of these curious peculiarities which experiment reveals.

Another difference, likewise very important, between the luminous and the sonorous vibrations, which also points out how little analogous can be the constitutions of the media which transmit the vibrations, appears in the phenomena of dispersion. The speed of propagation, which, as we have seen when discussing the measurement of the velocity of sound, depends very little on the musical note, is not at all the same in the case of the various radiations which can be propagated in the same substance. The index of refraction varies with the duration of the period, or, if you will, with the length of wave *in vacuo* which is proportioned to this duration, since *in vacuo* the speed of propagation is entirely the same for all vibrations.

Cauchy was the first to propose a theory on which other attempts have been modelled; for example, the very interesting and simple one of Briot. This last-named supposed that the luminous vibration could not perceptibly drag with it the molecular material of the medium across which it is propagated, but that matter, nevertheless, reacts on the ether with an intensity proportional to the elongation, in such a manner as tends to bring it back to its position of equilibrium. With this simple hypothesis we can fairly well interpret the phenomena of the dispersion of light in the case of transparent substances; but far from well, as M. Carvallo has noted in some extremely careful experiments, the dispersion of the infra-red spectrum, and not at all the peculiarities presented by absorbent substances.

M. Boussinesq arrives at almost similar results, by attributing dispersion, on the other hand, to the partial dragging along of ponderable matter and to its action on the ether. By combining, in a measure, as was subsequently done by M. Boussinesq, the two hypotheses, formulas can be established far better in accord with all the known facts.

These facts are somewhat complex. It was at first thought that the index always varied in inverse ratio to the wave-length, but numerous substances have been discovered which present the phenomenon of abnormal dispersion—that is to say, substances in which certain radiations are propagated, on the contrary, the more quickly the shorter their period. This is the case with gases themselves, as demonstrated, for example, by a very elegant experiment of M. Becquerel on the dispersion of the vapour of sodium. Moreover, it may happen that yet more complications may be met with, as no substance is transparent for the whole extent of the spectrum. In the case of certain radiations the speed of propagation becomes nil, and the index shows sometimes a maximum and sometimes a minimum. All those phenomena are in close relation with those of absorption.

It is, perhaps, the formula proposed by Helmholtz which best accounts for all these peculiarities. Helmholtz came to establish this formula by supposing that there is a kind of friction between the ether and matter, which, like that exercised on a pendulum, here produces a double effect, changing, on the one hand, the duration of this oscillation, and, on the other, gradually damping it. He further supposed that ponderable matter is acted on by elastic forces. The theory of Helmholtz has the great advantage of representing, not only the phenomena of dispersion, but also, as M. Carvallo has pointed out, the laws of rotatory polarization, its dispersion and other phenomena, among them the dichroism of the rotatory media discovered by M. Cotton.

In the establishment of these theories, the language of ordinary optics has always been employed. The phenomena are looked upon as due to mechanical deformations or to movements governed by certain forces. The electromagnetic theory leads, as we have seen, to the employment of other images. M.H. Poincaré, and, after him, Helmholtz, have both proposed electromagnetic theories of dispersion. On examining things closely, it will be found that there are not, in truth, in the two ways of regarding the problem, two equivalent translations of exterior reality. The electrical theory gives us to understand, much better than the mechanical one, that *in vacuo* the dispersion ought to be strictly null, and this absence of dispersion appears to be confirmed with extraordinary precision by astronomical observations. Thus the observation, often repeated, and at different times of year, proves that in the case of the star Algol, the light of which takes at least four years to reach us, no sensible difference in coloration accompanies the changes in brilliancy.

§ 2. THE THEORY OF LORENTZ

Purely mechanical considerations have therefore failed to give an entirely satisfactory interpretation of the phenomena in which even the simplest relations between matter and the ether appear. They would, evidently, be still more insufficient if used to explain certain effects produced on matter by light, which could not, without grave difficulties, be attributed to movement; for instance, the phenomena of electrification under the influence of certain radiations, or, again, chemical reactions such as photographic impressions.

The problem had to be approached by another road. The electromagnetic theory was a step in advance, but it comes to a standstill, so to speak, at the moment when the ether penetrates into matter. If we wish to go deeper into the inwardness of the phenomena, we must follow, for example, Professor Lorentz or Dr Larmor, and look with them for a mode of representation which appears, besides, to be a natural consequence of the fundamental ideas forming the basis of Hertz's experiments.

The moment we look upon a wave in the ether as an electromagnetic wave, a molecule which emits light ought to be considered as a kind of excitant. We are thus led to suppose that in each radiating molecule there are one or several electrified particles, animated with a to-and-fro movement round their positions of equilibrium, and these particles are certainly identical with those electrons the existence of which we have already admitted for so many other reasons.

In the simplest theory, we will imagine an electron which may be displaced from its position of equilibrium in all directions, and is, in this displacement, submitted to attractions which communicate to it a vibration like a pendulum. These movements are equivalent to tiny currents, and the mobile electron, when animated with a considerable velocity, must be sensitive to the action of the magnet which modifies the form of the trajectory and the value of the period. This almost direct consequence was perceived by Lorentz, and it led him to the new idea that radiations emitted by a body ought to be modified by the action of a strong electromagnet.

An experiment enabled this prevision to be verified. It was made, as is well known, as early as 1896 by Zeeman; and the discovery produced a legitimate sensation. When a flame is subjected to the action of a magnetic field, a brilliant line is decomposed in conditions more or less complex which an attentive study, however, allows us to define. According to whether the observation is made in a plane normal to the magnetic field or in the same direction, the line transforms itself into a triplet or doublet, and the new lines are polarized rectilinearly or circularly.

These are the precise phenomena which the calculation foretells: the analysis of the modifications undergone by the light supplies, moreover, valuable information on the electron itself. From the direction of the circular vibrations of the greatest frequency we can determine the sign of the electric charge in motion and we find it to be negative. But, further than this, from the variation of the period we can calculate the relation of the force acting on the electron to its material mass, and, in addition, the relation of the charge to the mass. We then find for this relation precisely that value which we have already met with so many times. Such a coincidence cannot be fortuitous, and we have the right to believe that the electron revealed by the luminous wave which emanates from it, is really the same as the one made known to us by the study of the cathode rays and of the radioactive substances.

However, the elementary theory does not suffice to interpret the complications which later experiments have revealed. The physicists most qualified to effect measurements in these delicate optical questions—M. Cornu, Mr Preston, M. Cotton, MM. Becquerel and Deslandres, M. Broca, Professor Michelson, and others—have pointed out some remarkable peculiarities. Thus in some cases the number of the component rays dissociated by the magnetic field may be very considerable.

The great modification brought to a radiation by the Zeeman effect may, besides, combine itself with other phenomena, and alter the light in a still more complicated manner. A pencil of polarized light, as demonstrated by Signori Macaluzo and Corbino, undergoes, in a magnetic field, modifications with regard to absorption and speed of propagation.

Some ingenious researches by M. Becquerel and M. Cotton have perfectly elucidated all these complications from an experimental point of view. It would not be impossible to link together all these phenomena without adopting the electronic hypothesis, by preserving the old optical equations as modified by the terms relating to the action of the magnetic field. This has actually been done in some very remarkable work by M. Voigt, but we may also, like Professor Lorentz, look for more general theories, in which the essential image of the electrons shall be preserved, and which will allow all the facts revealed by experiment to be included.

We are thus led to the supposition that there is not in the atom one vibrating electron only, but that there is to be found in it a dynamical system comprising several material points which may be subjected to varied movements. The neutral atom may therefore be considered as composed of an immovable principal portion positively charged, round which move, like satellites round a planet, several negative electrons of very inferior mass. This conclusion leads us to an interpretation in agreement with that which other phenomena have already suggested.

These electrons, which thus have a variable velocity, generate around themselves a transverse electromagnetic wave which is propagated with the velocity of light; for the charged particle becomes, as soon as it experiences a change of speed, the centre of a radiation. Thus is explained the phenomenon of the emission of radiations. In the same way, the movement of electrons may be excited or modified by the electrical forces which exist in any pencil of light they receive, and this pencil may yield up to them a part of the energy it is carrying. This is the phenomenon of absorption.

Professor Lorentz has not contented himself with thus explaining all the mechanism of the phenomena of emission and absorption. He has endeavoured to rediscover, by starting with the fundamental hypothesis, the quantitative laws discovered by thermodynamics. He succeeds in showing that, agreeably to the law of Kirchhoff, the relation between the emitting and the absorbing power must be independent of the special properties of the body under observation, and he thus again meets with the laws of Planck and of Wien: unfortunately the calculation can only be made in the case of great wave-lengths, and grave difficulties exist. Thus it cannot be very clearly explained why, by heating a body, the radiation is displaced towards the side of the short wave-lengths, or, if you will, why a body becomes luminous from the moment its temperature has reached a sufficiently high degree. On the other hand, by calculating the energy of the vibrating particles we are again led to attribute to these particles the same constitution as that of the electrons.

It is in the same way possible, as Professor Lorentz has shown, to give a very satisfactory explanation of the thermo-electric phenomena by supposing that the number of liberated electrons which exist in a given metal at a given temperature has a determined value varying with each metal, and is, in the case of each body, a function of the temperature. The formula obtained, which is based on these hypotheses, agrees completely with the classic results of Clausius and of Lord Kelvin. Finally, if we recollect that the phenomena of electric and calorific conductivity are perfectly interpreted by the hypothesis of electrons, it will no longer be possible to contest the importance of a theory which allows us to group together in one synthesis so many facts of such diverse origins.

If we study the conditions under which a wave excited by an electron's variations in speed can be transmitted, they again bring us face to face, and generally, with the results pointed out by the ordinary electromagnetic theory. Certain peculiarities, however, are not absolutely the same. Thus the theory of Lorentz, as well as that of Maxwell, leads us to foresee that if an insulating mass be caused to move in a magnetic field normally to its lines of force, a displacement will be produced in this mass analogous to that of which Faraday and Maxwell admitted the existence in the dielectric of a charged condenser. But M.H. Poincaré has pointed out that, according as we adopt one or other of these authors' points of view, so the value of the displacement differs. This remark is very important, for it may lead to an experiment which would enable us to make a definite choice between the two theories.

To obtain the displacement estimated according to Lorentz, we must multiply the displacement calculated according to Hertz by a factor representing the relation between the difference of the specific inductive capacities of the dielectric and of a vacuum, and the first of these powers. If therefore we take as dielectric the air of which the specific inductive capacity is perceptibly the same as that of a vacuum, the displacement, according to the idea of Lorentz, will be null; while, on the contrary, according to Hertz, it will have a finite value. M. Blondlot has made the experiment. He sent a current of air into a condenser placed in a magnetic field, and was never able to notice the slightest trace of electrification. No displacement, therefore, is effected in the dielectric. The experiment being a negative one, is evidently less convincing than one giving a positive result, but it furnishes a very powerful argument in favour of the theory of Lorentz.

This theory, therefore, appears very seductive, yet it still raises objections on the part of those who oppose to it the principles of ordinary mechanics. If we consider, for instance, a radiation emitted by an electron belonging to one material body, but absorbed by another electron in another body, we perceive immediately that, the propagation not being instantaneous, there can be no compensation between the action and the reaction, which are not simultaneous; and the principle of Newton thus seems to be attacked. In order to preserve its integrity, it has to be admitted that the movements in the two material substances are compensated by that of the ether which separates these substances; but this conception, although in tolerable agreement with the hypothesis that the ether and matter are not of different essence, involves, on a closer examination, suppositions hardly satisfactory as to the nature of movements in the ether.

For a long time physicists have admitted that the ether as a whole must be considered as being immovable and capable of serving, so to speak, as a support for the axes of Galileo, in relation to which axes the principle of inertia is applicable,—or better still, as M. Painlevé has shown, they alone allow us to render obedience to the principle of causality.

But if it were so, we might apparently hope, by experiments in electromagnetism, to obtain absolute motion, and to place in evidence the translation of the earth relatively to the ether. But all the researches attempted by the most ingenious physicists towards this end have always failed, and this tends towards the idea held by many geometricians that these negative results are not due to imperfections in the experiments, but have a deep and general cause. Now Lorentz has endeavoured to find the conditions in which the electromagnetic theory proposed by him might agree with the postulate of the complete impossibility of determining absolute motion. It is necessary, in order to realise this concord, to imagine that a mobile system contracts very slightly in the direction of its translation to a degree proportioned to the square of the ratio of the velocity of transport to that of light. The electrons themselves do not escape this contraction, although the observer, since he participates in the same motion, naturally cannot notice it. Lorentz supposes, besides, that all forces, whatever their origin, are affected by a translation in the same way as electromagnetic forces. M. Langevin and M. H. Poincaré have studied this same question and have noted with precision various delicate consequences of it. The singularity of the hypotheses which we are thus led to construct in no way constitutes an argument against the theory of Lorentz; but it has, we must acknowledge, discouraged some of the more timid partisans of this theory.[48]

§ 3. THE MASS OF ELECTRONS

Other conceptions, bolder still, are suggested by the results of certain interesting experiments. The electron affords us the possibility of considering inertia and mass to be no longer a fundamental notion, but a consequence of the electromagnetic phenomena.

Professor J.J. Thomson was the first to have the clear idea that a part, at least, of the inertia of an electrified body is due to its electric charge. This idea was taken up and precisely stated by Professor Max Abraham, who, for the first time, was led to regard seriously the seemingly paradoxical notion of mass as a function of velocity. Consider a small particle bearing a given electric charge, and let us suppose that this particle moves through the ether. It is, as we know, equivalent to a current proportional to its velocity, and it therefore creates a magnetic field the intensity of which is likewise proportional to its velocity: to set it in motion, therefore, there must be communicated to it over and above the expenditure corresponding to the acquisition of its ordinary kinetic energy, a quantity of energy proportional to the square of its velocity. Everything, therefore, takes place as if, by the fact of electrification, its capacity for kinetic energy and its material mass had been increased by a certain constant quantity. To the ordinary mass may be added, if you will, an electromagnetic mass.

This is the state of things so long as the speed of the translation of the particle is not very great, but they are no longer quite the same when this particle is animated with a movement whose rapidity becomes comparable to that with which light is propagated.

The magnetic field created is then no longer a field in repose, but its energy depends, in a complicated manner, on the velocity, and the apparent increase in the mass of the particle itself becomes a function of the velocity. More than this, this increase may not be the same for the same velocity, but varies according to whether the acceleration is parallel with or perpendicular to the direction of this velocity. In other words, there seems to be a longitudinal; and a transversal mass which need not be the same.

All these results would persist even if the material mass were very small relatively to the electromagnetic mass; and the electron possesses some inertia even if its ordinary mass becomes slighter and slighter. The apparent mass, it can be easily shown, increases indefinitely when the velocity with which the electrified particle is animated tends towards the velocity of light, and thus the work necessary to communicate such a velocity to an electron would be infinite. It is in consequence impossible that the speed of an electron, in relation to the ether, can ever exceed, or even permanently attain to, 300,000 kilometres per second.

All the facts thus predicted by the theory are confirmed by experiment. There is no known process which permits the direct measurement of the mass of an electron, but it is possible, as we have seen, to measure simultaneously its velocity and the relation of the electric charge to its mass. In the case of the cathode rays emitted by radium, these measurements are particularly interesting, for the reason that the rays which compose a pencil of cathode rays are animated by very different speeds, as is shown by the size of the stain produced on a photographic plate by a pencil of them at first very constricted and subsequently dispersed by the action of an electric or magnetic field. Professor Kaufmann has effected some very careful experiments by a method he terms the method of crossed spectra, which consists in superposing the deviations produced by a magnetic and an electric field respectively acting in directions at right angles one to another. He has thus been enabled by working *in vacuo* to register the very different velocities which, starting in the case of certain rays from about seven-tenths of the velocity of light, attain in other cases to ninety-five hundredths of it.

It is thus noted that the ratio of charge to mass—which for ordinary speeds is constant and equal to that already found by so many experiments—diminishes slowly at first, and then very rapidly when the velocity of the ray increases and approaches that of light. If we represent this variation by a curve, the shape of this curve inclines us to think that the ratio tends toward zero when the velocity tends towards that of light.

All the earlier experiments have led us to consider that the electric charge was the same for all electrons, and it can hardly be conceived that this charge can vary with the velocity. For in order that the relation, of which one of the terms remains fixed, should vary, the other term necessarily cannot remain constant. The experiments of Professor Kaufmann, therefore, confirm the previsions of Max Abraham's theory: the mass depends on the velocity, and increases indefinitely in proportion as this velocity approaches that of light. These experiments, moreover, allow the numerical results of the calculation to be compared with the values measured. This very satisfactory comparison shows that the apparent total mass is sensibly equal to the electromagnetic mass; the material mass of the electron is therefore nil, and the whole of its mass is electromagnetic.

Thus the electron must be looked upon as a simple electric charge devoid of matter. Previous examination has led us to attribute to it a mass a thousand times less that that of the atom of hydrogen, and a more attentive study shows that this mass was fictitious. The electromagnetic phenomena which are produced when the electron is set in motion or a change effected in its velocity, simply have the effect, as it were, of simulating inertia, and it is the inertia due to the charge which has caused us to be thus deluded.

The electron is therefore simply a small volume determined at a point in the ether, and possessing special properties; [49] this point is propagated with a velocity which cannot exceed that of light. When this velocity is constant, the electron creates around it in its passage an electric and a magnetic field; round this electrified centre there exists a kind of wake, which follows it through the ether and does not become modified so long as the velocity remains invariable. If other electrons follow the first within a wire, their passage along the wire will be what is called an electric current.

When the electron is subjected to an acceleration, a transverse wave is produced, and an electromagnetic radiation is generated, of which the character may naturally change with the manner in which the speed varies. If the electron has a sufficiently rapid periodical movement, this wave is a light wave; while if the electron stops suddenly, a kind of pulsation is transmitted through the ether, and thus we obtain Röntgen rays.

§ 4. NEW VIEWS ON THE CONSTITUTION OF THE ETHER AND OF MATTER

New and valuable information is thus afforded us regarding the properties of the ether, but will this enable us to construct a material representation of this medium which fills the universe, and so to solve a problem which has baffled, as we have seen, the prolonged efforts of our predecessors?

Certain scholars seem to have cherished this hope. Dr. Larmor in particular, as we have seen, has proposed a most ingenious image, but one which is manifestly insufficient. The present tendency of physicists rather tends to the opposite view; since they consider matter as a very complex object, regarding which we wrongly imagine ourselves to be well informed because we are so much accustomed to it, and its singular properties end by seeming natural to us. But in all probability the ether is, in its objective reality, much more simple, and has a better right to be considered as fundamental.

We cannot therefore, without being very illogical, define the ether by material properties, and it is useless labour, condemned beforehand to sterility, to endeavour to determine it by other qualities than those of which experiment gives us direct and exact knowledge.

The ether is defined when we know, in all its points, and in magnitude and in direction, the two fields, electric and magnetic, which may exist in it. These two fields may vary; we speak from habit of a movement propagated in the ether, but the phenomenon within the reach of experiment is the propagation of these variations.

Since the electrons, considered as a modification of the ether symmetrically distributed round a point, perfectly counterfeit that inertia which is the fundamental property of matter, it becomes very tempting to suppose that matter itself is composed of a more or less complex assemblage of electrified centres in motion.

This complexity is, in general, very great, as is demonstrated by the examination of the luminous spectra produced by the atoms, and it is precisely because of the compensations produced between the different movements that the essential properties of matter—the law of the conservation of inertia, for example—are not contrary to the hypothesis.

The forces of cohesion thus would be due to the mutual attractions which occur in the electric and magnetic fields produced in the interior of bodies; and it is even conceivable that there may be produced, under the influence of these actions, a tendency to determine orientation, that is to say, that a reason can be seen why matter may be crystallised.[50]

All the experiments effected on the conductivity of gases or metals, and on the radiations of active bodies, have induced us to regard the atom as being constituted by a positively charged centre having practically the same magnitude as the atom itself, round which the electrons gravitate; and it might evidently be supposed that this positive centre itself preserves the fundamental characteristics of matter, and that it is the electrons alone which no longer possess any but electromagnetic mass.

We have but little information concerning these positive particles, though they are met with in an isolated condition, as we have seen, in the canal rays or in the X rays.[51] It has not hitherto been possible to study them so successfully as the electrons themselves; but that their magnitude causes them to produce considerable perturbations in the bodies on which they fall is manifest by the secondary emissions which complicate and mask the primitive phenomenon. There are, however, strong reasons for thinking that these positive centres are not simple. Thus Professor Stark attributes to them, with experiments in proof of his opinion, the emission of the spectra of the rays in Geissler tubes, and the complexity of the spectrum discloses the complexity of the centre. Besides, certain peculiarities in the conductivity of metals cannot be explained without a supposition of this kind. So that the atom, deprived of the cathode corpuscle, would be still liable to decomposition into elements analogous to electrons and positively charged. Consequently nothing prevents us supposing that this centre likewise simulates inertia by its electromagnetic properties, and is but a condition localised in the ether.

However this may be, the edifice thus constructed, being composed of electrons in periodical motion, necessarily grows old. The electrons become subject to accelerations which produce a radiation towards the exterior of the atom; and certain of them may leave the body, while the primitive stability is, in the end, no longer assured, and a new arrangement tends to be formed. Matter thus seems to us to undergo those transformations of which the radio-active bodies have given us such remarkable examples.

We have already had, in fragments, these views on the constitution of matter; a deeper study of the electron thus enables us to take up a position from which we obtain a sharp, clear, and comprehensive grasp of the whole and a glimpse of indefinite horizons.

It would be advantageous, however, in order to strengthen this position, that a few objections which still menace it should be removed. The instability of the electron is not yet sufficiently demonstrated. How is it that its charge does not waste itself away, and what bonds assure the permanence of its constitution?

On the other hand, the phenomena of gravitation remain a mystery. Lorentz has endeavoured to build up a theory in which he explains attraction by supposing that two charges of similar sign repel each other in a slightly less degree than that in which two charges, equal but of contrary sign, attract each other, the difference being, however, according to the calculation, much too small to be directly observed. He has also sought to explain gravitation by connecting it with the pressures which may be produced on bodies by the vibratory movements which form very penetrating rays. Recently M. Sutherland has imagined that attraction is due to the difference of action in the convection currents produced by the positive and negative corpuscles which constitute the atoms of the stars, and are carried along by the astronomical motions. But these hypotheses remain rather vague, and many authors think, like M. Langevin, that gravitation must result from some mode of activity of the ether totally different from the electromagnetic mode.

CHAPTER XI
THE FUTURE OF PHYSICS

It would doubtless be exceedingly rash, and certainly very presumptuous, to seek to predict the future which may be reserved for physics. The rôle of prophet is not a scientific one, and the most firmly established previsions of to-day may be overthrown by the reality of to-morrow.

Nevertheless, the physicist does not shun an extrapolation of some little scope when it is not too far from the realms of experiment; the knowledge of the evolution accomplished of late years authorises a few suppositions as to the direction in which progress may continue.

The reader who has deigned to follow me in the rapid excursion we have just made through the domain of the science of Nature, will doubtless bring back with him from his short journey the general impression that the ancient limits to which the classic treatises still delight in restricting the divers chapters of physics, are trampled down in all directions.

The fine straight roads traced out by the masters of the last century, and enlarged and levelled by the labour of such numbers of workmen, are now joined together by a crowd of small paths which furrow the field of physics. It is not only because they cover regions as yet little explored where discoveries are more abundant and more easy, that these cross-cuts are so frequent, but also because a higher hope guides the seekers who engage in these new routes.

In spite of the repeated failures which have followed the numerous attempts of past times, the idea has not been abandoned of one day conquering the supreme principle which must command the whole of physics.

Some physicists, no doubt, think such a synthesis to be impossible of realisation, and that Nature is infinitely complex; but, notwithstanding all the reserves they may make, from the philosophical point of view, as to the legitimacy of the process, they do not hesitate to construct general hypotheses which, in default of complete mental satisfaction, at least furnish them with a highly convenient means of grouping an immense number of facts till then scattered abroad.

Their error, if error there be, is beneficial, for it is one of those that Kant would have classed among the fruitful illusions which engender the indefinite progress of science and lead to great and important co-ordinations.

It is, naturally, by the study of the relations existing between phenomena apparently of very different orders that there can be any hope of reaching the goal; and it is this which justifies the peculiar interest accorded to researches effected in the debatable land between domains hitherto considered as separate.

Among all the theories lately proposed, that of the ions has taken a preponderant place; ill understood at first by some, appearing somewhat singular, and in any case useless, to others, it met at its inception, in France at least, with only very moderate favour.

To-day things have greatly changed, and those even who ignored it have been seduced by the curious way in which it adapts itself to the interpretation of the most recent experiments on very different subjects. A very natural reaction has set in; and I might almost say that a question of fashion has led to some exaggerations.

The electron has conquered physics, and many adore the new idol rather blindly. Certainly we can only bow before an hypothesis which enables us to group in the same synthesis all the discoveries on electric discharges and on radioactive substances, and which leads to a satisfactory theory of optics and of electricity; while by the intermediary of radiating heat it seems likely to embrace shortly the principles of thermodynamics also. Certainly one must admire the power of a creed which penetrates also into the domain of mechanics and furnishes a simple representation of the essential properties of matter; but it is right not to lose sight of the fact that an image may be a well-founded appearance, but may not be capable of being exactly superposed on the objective reality.

The conception of the atom of electricity, the foundation of the material atoms, evidently enables us to penetrate further into Nature's secrets than our predecessors; but we must not be satisfied with words, and the mystery is not solved when, by a legitimate artifice, the difficulty has simply been thrust further back. We have transferred to an element ever smaller and smaller those physical qualities which in antiquity were attributed to the whole of a substance; and then we shifted them later to those chemical atoms which, united together, constitute this whole. To-day we pass them on to the electrons which compose these atoms. The indivisible is thus rendered, in a way, smaller and smaller, but we are still unacquainted with what its substance may be. The notion of an electric charge which we substitute for that of a material mass will permit phenomena to be united which we thought separate, but it cannot be considered a definite explanation, or as the term at which science must stop. It is probable, however, that for a few years still physics will not travel beyond it. The present hypothesis suffices for grouping known facts, and it will doubtless enable many more to be foreseen, while new successes will further increase its possessions.

Then the day will arrive when, like all those which have shone before it, this seductive hypothesis will lead to more errors than discoveries. It will, however, have been improved, and it will have become a very vast and very complete edifice which some will not willingly abandon; for those who have made to themselves a comfortable dwelling-place on the ruins of ancient monuments are often too loth to leave it.

In that day the searchers who were in the van of the march after truth will be caught up and even passed by others who will have followed a longer, but perhaps surer road. We also have seen at work those prudent physicists who dreaded too daring creeds, and who sought only to collect all the documentary evidence possible, or only took for their guide a few principles which were to them a simple generalisation of facts established by experiments; and we have been able to prove that they also were effecting good and highly useful work.

Neither the former nor the latter, however, carry out their work in an isolated way, and it should be noted that most of the remarkable results of these last years are due to physicists who have known how to combine their efforts and to direct their activity towards a common object, while perhaps it may not be useless to observe also that progress has been in proportion to the material resources of our laboratories.

It is probable that in the future, as in the past, the greatest discoveries, those which will suddenly reveal totally unknown regions, and open up entirely new horizons, will be made by a few scholars of genius who will carry on their patient labour in solitary meditation, and who, in order to verify their boldest conceptions, will no doubt content themselves with the most simple and least costly experimental apparatus. Yet for their discoveries to yield their full harvest, for the domain to be systematically worked and desirable results obtained, there will be more and more required the association of willing minds, the solidarity of intelligent scholars, and it will be also necessary for these last to have at their disposal the most delicate as well as the most powerful instruments. These are conditions paramount at the present day for continuous progress in experimental science.

If, as has already happened, unfortunately, in the history of science, these conditions are not complied with; if the freedoms of the workers are trammelled, their unity disturbed, and if material facilities are too parsimoniously afforded them,—evolution, at present so rapid, may be retarded, and those retrogressions which, by-the-by, have been known in all evolutions, may occur, although even then hope in the future would not be abolished for ever.

There are no limits to progress, and the field of our investigations has no boundaries. Evolution will continue with invincible force. What we to-day call the unknowable, will retreat further and further before science, which will never stay her onward march. Thus physics will give greater and increasing satisfaction to the mind by furnishing new interpretations of phenomena; but it will accomplish, for the whole of society, more valuable work still, by rendering, by the improvements it suggests, life every day more easy and more agreeable, and by providing mankind with weapons against the hostile forces of Nature.

FOOTNOTES

[1] *I.e.*, the time-curve.—ED.

[2] The author seems to refer to the fact that in the standard metre, the measurement is taken from the central one of three marks at each end of the bar. The transverse section of the bar is an X, and the reading is made by a microscope.—ED.

[3] *I.e.* 1/2000 of a millimetre.—ED.

[4] These are the magnitudes and units adopted at the International Congress of Electricians in 1904. For their definition and explanation, see Demanet, *Notes de Physique Expérimentale* (Louvain, 1905), t. iv. p. 8.—ED.

[5] "Nothing is created; nothing is lost"—ED.

[6] By isothermal diagram is meant the pattern or complex formed when the isothermal lines are arranged in curves of which the pressure is the ordinate and the volume the abscissa.—ED.

[7] Mr Preston thus puts it: "The law [of corresponding states] seems to be not quite, but very nearly true for these substances [*i.e.* the halogen derivatives of benzene]; but in the case of the other substances examined, the majority of these generalizations were either only roughly true or altogether departed from" (*Theory of Heat*, London, 1904, p. 514.)—ED.

[8] Methode avec retour en arriere.—ED

[9] Professor Soddy, in a paper read before the Royal Society on the 15th November 1906, warns experimenters against vacua created by charcoal cooled in liquid air (the method referred-to in the text), unless as much of the air as possible is first removed with a pump and replaced by some argon-free gas. According to him, neither helium nor argon is absorbed by charcoal. By

the use of electrically-heated calcium, he claims to have produced an almost perfect vacuum.—ED.

[10] Another view, viz. that these inert gases are a kind of waste product of radioactive changes, is also gaining ground. The discovery of the radioactive mineral malacone, which gives off both helium and argon, goes to support this. See Messrs Ketchin and Winterson's paper on the subject at the Chemical Society, 18th October 1906.—ED.

[11] M. Poincaré is here in error. Helium has never been liquefied.—ED.

[12] Professor Quincke's last hypothesis is that all liquids on solidifying pass through a stage intermediate between solid and liquid, in which they form what he calls "foam-cells," and assume a viscous structure resembling that of jelly. See *Proc. Roy. Soc. A.*, 23rd July 1906.—ED.

[13] The metal known as "invar."—ED.

[14] The "second principle" referred to has been thus enunciated: "In every engine that produces work there is a fall of temperature, and the maximum output of a perfect engine—*i.e.* the ratio between the heat consumed in work and the heat supplied—depends only on the extreme temperatures between which the fluid is evolved."—Demanet, *Notes de Physique Expérimentale*, Louvain, 1905, fasc. 2, p. 147. Clausius put it in a negative form, as thus: No engine can of itself, without the aid of external agency, transfer heat from a body at low temperature to a body at a high temperature. *Cf.* Ganot's *Physics*, 17th English edition, § 508.—ED.

[15] See next note.—ED.

[16] M. Stephane Leduc, Professor of Biology of Nantes, has made many experiments in this connection, and the artificial cells exhibited by him to the Association française pour l'avancement des Sciences, at their meeting at Grenoble in 1904 and reproduced in their "Actes," are particularly noteworthy.—ED.

[17] That is, without receiving or emitting any heat.—ED.

[18] Dissociation must be distinguished from decomposition, which is what occurs when the whole of a particle (compound, molecule, atom, etc.) breaks up into its component parts. In dissociation the breaking up is only partial, and the resultant consists of a mixture of decomposed and undecomposed parts. See Ganot's Physics, 17th English edition, § 395, for examples.—ED.

[19] The valency or atomicity of an element may be defined as the power it possesses of entering into compounds in a certain fixed proportion. As hydrogen is generally taken as the standard, in practice the valency of an

atom is the number of hydrogen atoms it will combine with or replace. Thus chlorine and the rest of the halogens, the atoms of which combine with one atom of hydrogen, are called univalent, oxygen a bivalent element, and so on.—ED.

[20] Since this was written, however, men of science have become less unanimous than they formerly were on this point. The veteran chemist Professor Mendeléeff has given reasons for thinking that the ether is an inert gas with an atomic weight a million times less than that of hydrogen, and a velocity of 2250 kilometres per second (*Principles of Chemistry*, Eng. ed., 1905, vol. ii. p. 526). On the other hand, the well-known physicist Dr A.H. Bucherer, speaking at the Naturforscherversammlung, held at Stuttgart in 1906, declared his disbelief in the existence of the ether, which he thought could not be reconciled at once with the Maxwellian theory and the known facts.—ED.

[21] A natural chlorate of potassium, generally of volcanic origin.—ED.

[22] That is to say, he reflected the beam of polarized light by a mirror placed at that angle. See Turpain, *Leçons élementaires de Physique*, t. ii. p. 311, for details of the experiment.—ED.

[23] It will no doubt be a shock to those whom Professor Henry Armstrong has lately called the "mathematically-minded" to find a member of the Poincaré family speaking disrespectfully of the science they have done so much to illustrate. One may perhaps compare the expression in the text with M. Henri Poincaré's remark in his last allocution to the Académie des Sciences, that "Mathematics are sometimes a nuisance, and even a danger, when they induce us to affirm more than we know" (*Comptes-rendus*, 17th December 1906).

[24] See footnote 3.

[25] *I.e.* 10,000 metres.—ED.

[26] By this M. Poincaré appears to mean a radiometer in which the vanes are not entirely free to move as in the radiometer of Crookes but are suspended by one or two threads as in the instrument devised by Professor Poynting.—ED.

[27] See especially the experiments of Professor E. Marx (Vienna), *Annalen der Physik*, vol. xx. (No. 9 of 1906), pp. 677 *et seq.*, which seem conclusive on this point.—ED.

[28] M. Sagnac (*Le Radium*, Jan. 1906, p. 14), following perhaps Professors Elster and Geitel, has lately taken up this idea anew.—ED.

[29] At least, so long as it is not introduced between the two coatings of a condenser having a difference of potential sufficient to overcome what M. Bouty calls its dielectric cohesion. We leave on one side this phenomenon, regarding which M. Bouty has arrived at extremely important results by a very remarkable series of experiments; but this question rightly belongs to a special study of electrical phenomena which is not yet written.

[30] A full account of these experiments, which were executed at the Cavendish Laboratory, is to be found in *Philosophical Transactions*, A., vol. cxcv. (1901), pp. 193 *et seq.*—ED.

[31] The whole of this argument is brilliantly set forth by Professor Lorentz in a lecture delivered to the Electrotechnikerverein at Berlin in December 1904, and reprinted, with additions, in the *Archives Néerlandaises* of 1906.—ED.

[32] In his work on *L'Évolution de la Matière*, M. Gustave Le Bon recalls that in 1897 he published several notes in the Académie des Sciences, in which he asserted that the properties of uranium were only a particular case of a very general law, and that the radiations emitted did not polarize, and were akin by their properties to the X rays.

[33] Polonium has now been shown to be no new element, but one of the transformation products of radium. Radium itself is also thought to be derived in some manner, not yet ascertained, from uranium. The same is the case with actinium, which is said to come in the long run from uranium, but not so directly as does radium. All this is described in Professor Rutherford's *Radioactive Transformations* (London, 1906).—ED.

[34] This is admitted by Professor Rutherford (*Radio-Activity*, Camb., 1904, p. 141) and Professor Soddy (*Radio-Activity*, London, 1904, p. 66). Neither Mr Whetham, in his Recent *Development of Physical Science* (London, 1904) nor the Hon. R.J. Strutt in *The Becquerel Rays* (London, same date), both of whom deal with the historical side of the subject, seem to have noticed the fact.—ED.

[35] It has now been shown that polonium when freshly separated emits beta rays also; see Dr Logeman's paper in *Proceedings of the Royal Society*, A., 6th September 1906.—ED.

[36] According to Professor Rutherford, in 3.77 days.—ED

[37] Professor Rutherford has lately stated that uranium may possibly produce an emanation, but that its rate of decay must be too swift for its presence to be verified (see *Radioactive Transformations*, p. 161).—ED.

[38] An actinium X was also discovered by Professor Giesel (*Jahrbuch d. Radioaktivitat*, i. p. 358, 1904). Since the above was written, another product has been found to intervene between the X substance and the emanation in the case of actinium and thorium. They have been named radio-actinium and radio-thorium respectively.—ED.

[39] Such a table is given on p. 169 of Rutherford's *Radioactive Transformations*.—ED.

[40] This opinion, no doubt formed when Sir William Ramsay's discovery of the formation of helium from the radium emanation was first made known, is now less tenable. The latest theory is that the alpha particle is in fact an atom of helium, and that the final transformation product of radium and the other radioactive substances is lead. Cf. Rutherford, *op. cit. passim.*—ED.

[41] See *Radioactive Transformations* (p. 251). Professor Rutherford says that "each of the alpha ray products present in one gram of radium product (*sic*) expels 6.2×10^{10} alpha particles per second." He also remarks on "the experimental difficulty of accurately determining the number of alpha particles expelled from radium per second."—ED.

[42] See Rutherford, *op. cit.* p. 150.—ED.

[43] This view of the case has been made very clear by M. Gustave le Bon in *L'Évolution de la Matière* (Paris, 1906). See especially pp. 36-52, where the amount of the supposed intra-atomic energy is calculated.—ED.

[44] This is the main contention of M. Gustave Le Bon in his work last quoted.—ED.

[45] See last note.—ED.

[46] In reality M. Sagnac operated in the converse manner. He took two equal *weights* of a salt of radium and a salt of barium, which he made oscillate one after the other in a torsion balance. Had the durations of oscillation been different, it might be concluded that the mechanical mass is not the same for radium as for barium.

[47] Many theories as to the cause of the lines and bands of the spectrum have been put forward since this was written, among which that of Professor Stark (for which see *Physikalische Zeitschrift* for 1906, *passim*) is perhaps the most advanced. That of M. Jean Becquerel, which would attribute it to the vibration within the atom of both negative and positive electrons, also deserves notice. A popular account of this is given in the *Athenæum* of 20th April 1907.—ED.

[48] An objection not here noticed has lately been formulated with much frankness by Professor Lorentz himself. It is one of the pillars of his theory

that only the negative electrons move when an electric current passes through a metal, and that the positive electrons (if any such there be) remain motionless. Yet in the experiment known as Hall's, the current is deflected by the magnetic field to one side of the strip in certain metals, and to the opposite side in others. This seems to show that in certain cases the positive electrons move instead of the negative, and Professor Lorentz confesses that up to the present he can find no valid argument against this. See *Archives Néerlandaises* 1906, parts 1 and 2.—ED.

[49] This cannot be said to be yet completely proved. *Cf.* Sir Oliver Lodge, *Electrons*, London, 1906, p. 200.—ED.

[50] The reader should, however, be warned that a theory has lately been put forth which attempts to account for crystallisation on purely mechanical grounds. See Messrs Barlow and Pope's "Development of the Atomic Theory" in the *Transactions of the Chemical Society*, 1906.—ED.

[51] There is much reason for thinking that the canal rays do not contain positive particles alone, but are accompanied by negative electrons of slow velocity. The X rays are thought, as has been said above, to contain neither negative nor positive particles, but to be merely pulses in the ether.—ED.

Tesla coil web.

Tesla Laboratory.

Tesla at age 23

In 1885, at age 29

In 1895, at age 39

In 1915, at age 59

In 1920, at age 64

At a press conference at
the Hotel New Yorker, July 10, 1935,
his 79th birthday

Mark Twain observing experiments inside N. Tesla's Laboratory.

Archimedes

Sir Isaac Newton

LEONARDO DAVINCI

RELATIVITY:
THE SPECIAL AND GENERAL THEORY

BY ALBERT EINSTEIN
Translated By Robert W. Lawson

Publisher: Methuen & Co Ltd
First Published: December, 1916
Revised Edition
©1916.

Written: 1916
(This Revised Edition: 1924, 2012)
First Published: December, 1916
Translated: Robert W. Lawson (Authorized Translation)

Albert Einstein

Mark Twain inside N. Tesla's Laboratory

CONTENTS

Note: The fifth Appendix was added by Einstein at the time of the fifteenth reprinting of this book; and as a result is still under copyright restrictions so cannot be added without the permission of the publisher.

PREFACE

(December, 1916)

The present book is intended, as far as possible, to give an exact insight into the theory of Relativity to those readers who, from a general scientific and philosophical point of view, are interested in the theory, but who are not conversant with the mathematical apparatus of theoretical physics. The work presumes a standard of education corresponding to that of a university matriculation examination, and, despite the shortness of the book, a fair amount of patience and force of will on the part of the reader. The author has spared himself no pains in his endeavour to present the main ideas in the simplest and most intelligible form, and on the whole, in the sequence and connection in which they actually originated. In the interest of clearness, it appeared to me inevitable that I should repeat myself frequently, without paying the slightest attention to the elegance of the presentation. I adhered scrupulously to the precept of that brilliant theoretical physicist L. Boltzmann, according to whom matters of elegance ought to be left to the tailor and to the cobbler. I make no pretence of having withheld from the reader difficulties which are inherent to the subject. On the other hand, I have purposely treated the empirical physical foundations of the theory in a "step-motherly" fashion, so that readers unfamiliar with physics may not feel like the wanderer who was unable to see the forest for the trees. May the book bring some one a few happy hours of suggestive thought!

December, 1916
A. EINSTEIN

PART I : THE SPECIAL
THEORY OF RELATIVITY

1. PHYSICAL MEANING OF GEOMETRICAL PROPOSITIONS

In your schooldays most of you who read this book made acquaintance with the noble building of Euclid's geometry, and you remember--perhaps with more respect than love--the magnificent structure, on the lofty staircase of which you were chased about for uncounted hours by conscientious teachers. By reason of our past experience, you would certainly regard everyone with disdain who should pronounce even the most out-of-the-way proposition of this science to be untrue. But perhaps this feeling of proud certainty would leave you immediately if someone were to ask you: "What, then, do you mean by the assertion that these propositions are true?" Let us proceed to give this question a little consideration.

Geometry sets out from certain conceptions such as "plane," "point," and "straight line," with which we are able to associate more or less definite ideas, and from certain simple propositions (axioms) which, in virtue of these ideas, we are inclined to accept as "true." Then, on the basis of a logical process, the justification of which we feel ourselves compelled to admit, all remaining propositions are shown to follow from those axioms, i.e. they are proven. A proposition is then correct ("true") when it has been derived in the recognised manner from the axioms. The question of "truth" of the individual geometrical propositions is thus reduced to one of the "truth" of the axioms. Now it has long been known that the last question is not only unanswerable by the methods of geometry, but that it is in itself entirely without meaning. We cannot ask whether it is true that only one straight line goes through two points. We can only say that Euclidean geometry deals with things called "straight lines," to each of which is ascribed the property of being uniquely determined by two points situated on it. The concept "true" does not tally with the assertions of pure geometry, because by the word "true" we are eventually in the habit of designating always the correspondence with a "real" object; geometry, however, is not concerned with the relation of the ideas involved in it to objects of experience, but only with the logical connection of these ideas among themselves.

It is not difficult to understand why, in spite of this, we feel constrained to call the propositions of geometry "true." Geometrical ideas correspond to more or less exact objects in nature, and these last are undoubtedly the exclusive cause of the genesis of those ideas. Geometry ought to refrain from such a course, in order to give to its structure the largest possible logical unity. The practice, for example, of seeing in a "distance" two marked positions on a practically rigid body is something which is lodged deeply in our habit of thought. We are accustomed further to regard three points as being situated on a straight line, if their apparent positions can be made to coincide for observation with one eye, under suitable choice of our place of observation.

If, in pursuance of our habit of thought, we now supplement the propositions of Euclidean geometry by the single proposition that two points on a practically rigid body always correspond to the same distance (line-interval), independently of any changes in position to which we may subject the body, the propositions of Euclidean geometry then resolve themselves into propositions on the possible relative position of practically rigid bodies.[1] Geometry which has been supplemented in this way is then to be treated as a branch of physics. We can now legitimately ask as to the "truth" of geometrical propositions interpreted in this way, since we are justified in asking whether these propositions are satisfied for those real things we have associated with the geometrical ideas. In less exact terms we can express this by saying that by the "truth" of a geometrical proposition in this sense we understand its validity for a construction with rule and compasses.

Of course the conviction of the "truth" of geometrical propositions in this sense is founded exclusively on rather incomplete experience. For the present we shall assume the "truth" of the geometrical propositions, then at a later stage (in the general theory of relativity) we shall see that this "truth" is limited, and we shall consider the extent of its limitation.

Notes

1) It follows that a natural object is associated also with a straight line. Three points A, B and C on a rigid body thus lie in a straight line when the points A and C being given, B is chosen such that the sum of the distances AB and BC is as short as possible. This incomplete suggestion will suffice for the present purpose.

2. THE SYSTEM OF CO-ORDINATES

On the basis of the physical interpretation of distance which has been indicated, we are also in a position to establish the distance between two points on a rigid body by means of measurements. For this purpose we require a " distance " (rod S) which is to be used once and for all, and which we employ as a standard measure. If, now, A and B are two points on a rigid body, we can construct the line joining them according to the rules of geometry ; then, starting from A, we can mark off the distance S time after time until we reach B. The number of these operations required is the numerical measure of the distance AB. This is the basis of all measurement of length.[2]

Every description of the scene of an event or of the position of an object in space is based on the specification of the point on a rigid body (body of reference) with which that event or object coincides. This applies not only to scientific description, but also to everyday life. If I analyse the place specification " Times Square, New York," [3] I arrive at the following result. The earth is the rigid body to which the specification of place refers; "Times Square, New York," is a well-defined point, to which a name has been assigned, and with which the event coincides in space. [4]

This primitive method of place specification deals only with places on the surface of rigid bodies, and is dependent on the existence of points on this surface which are distinguishable from each other. But we can free ourselves from both of these limitations without altering the nature of our specification of position. If, for instance, a cloud is hovering over Times Square, then we can determine its position relative to the surface of the earth by erecting a pole perpendicularly on the Square, so that it reaches the cloud. The length of the pole measured with the standard measuring-rod, combined with the specification of the position of the foot of the pole, supplies us with a complete place specification. On the basis of this illustration, we are able to see the manner in which a refinement of the conception of position has been developed.

(a) We imagine the rigid body, to which the place specification is referred, supplemented in such a manner that the object whose position we require is reached by. the completed rigid body.

(b) In locating the position of the object, we make use of a number (here the length of the pole measured with the measuring-rod) instead of designated points of reference.

(c) We speak of the height of the cloud even when the pole which reaches the cloud has not been erected. By means of optical observations of the cloud from different positions on the ground, and taking into account the properties of the propagation of light, we determine the length of the pole we should have required in order to reach the cloud.

From this consideration we see that it will be advantageous if, in the description of position, it should be possible by means of numerical measures to make ourselves independent of the existence of marked positions (possessing names) on the rigid body of reference. In the physics of measurement this is attained by the application of the Cartesian system of co-ordinates.

This consists of three plane surfaces perpendicular to each other and rigidly attached to a rigid body. Referred to a system of co-ordinates, the scene of any event will be determined (for the main part) by the specification of the lengths of the three perpendiculars or co-ordinates (x, y, z) which can be dropped from the scene of the event to those three plane surfaces. The lengths of these three perpendiculars can be determined by a series of manipulations with rigid measuring-rods performed according to the rules and methods laid down by Euclidean geometry.

In practice, the rigid surfaces which constitute the system of co-ordinates are generally not available ; furthermore, the magnitudes of the co-ordinates are not actually determined by constructions with rigid rods, but by indirect means. If the results of physics and astronomy are to maintain their clearness, the physical meaning of specifications of position must always be sought in accordance with the above considerations. [5)]

We thus obtain the following result: Every description of events in space involves the use of a rigid body to which such events have to be referred. The resulting relationship takes for granted that the laws of Euclidean geometry hold for "distances;" the "distance" being represented physically by means of the convention of two marks on a rigid body.

Notes

2) Here we have assumed that there is nothing left over i.e. that the measurement gives a whole number. This difficulty is got over by the use of divided measuring-rods, the introduction of which does not demand any fundamentally new method.

3) Einstein used "Potsdamer Platz, Berlin" in the original text. In the authorised translation this was supplemented with "Tranfalgar Square, London". We have changed this to "Times Square, New York", as this is the most well known/identifiable location to English speakers in the present day. [Note by the janitor.]

4) It is not necessary here to investigate further the significance of the expression "coincidence in space." This conception is sufficiently obvious to ensure that differences of opinion are scarcely likely to arise as to its applicability in practice.

5) A refinement and modification of these views does not become necessary until we come to deal with the general theory of relativity, treated in the second part of this book.

3. SPACE AND TIME IN CLASSICAL MECHANICS

The purpose of mechanics is to describe how bodies change their position in space with "time." I should load my conscience with grave sins against the sacred spirit of lucidity were I to formulate the aims of mechanics in this way, without serious reflection and detailed explanations. Let us proceed to disclose these sins.

It is not clear what is to be understood here by "position" and "space." I stand at the window of a railway carriage which is travelling uniformly, and drop a stone on the embankment, without throwing it. Then, disregarding the influence of the air resistance,

I see the stone descend in a straight line. A pedestrian who observes the misdeed from the footpath notices that the stone falls to earth in a parabolic curve. I now ask: Do the "positions" traversed by the stone lie "in reality" on a straight line or on a parabola? Moreover, what is meant here by motion "in space"? From the considerations of the previous section the answer is self-evident. In the first place we entirely shun the vague word "space," of which, we must honestly acknowledge, we cannot form the slightest conception, and we replace it by "motion relative to a practically rigid body of reference." The positions relative to the body of reference (railway carriage or embankment) have already been defined in detail in the preceding section. If instead of "body of reference" we insert "system of co-ordinates," which is a useful idea for mathematical description, we are in a position to say : The stone traverses a straight line relative to a system of co-ordinates rigidly attached to the carriage, but relative to a system of co-ordinates rigidly attached to the ground (embankment) it describes a parabola. With the aid of this example it is clearly seen that there is no such thing as an independently existing trajectory (lit. "path-curve"[6]), but only a trajectory relative to a particular body of reference.

In order to have a complete description of the motion, we must specify how the body alters its position with time ; i.e. for every point on the trajectory it must be stated at what time the body is situated there. These data must be supplemented by such a definition of time that, in virtue of this definition, these time-values can be regarded essentially as magnitudes (results of measurements) capable of observation. If we take our stand on the ground of classical mechanics, we can satisfy this requirement for our illustration in the following manner. We imagine two clocks of identical construction ; the man at the railway-carriage window is holding one of them, and the man on the footpath the other. Each of the observers determines the position on his own reference-body occupied by the stone at each tick of the clock he is holding in his hand. In this connection we have not taken account of the inaccuracy involved by the finiteness of the velocity of propagation of light. With this and with a second difficulty prevailing here we shall have to deal in detail later.

Notes

6) That is, a curve along which the body moves.

4. THE GALILEIAN SYSTEM OF CO-ORDINATES

As is well known, the fundamental law of the mechanics of Galilei-Newton, which is known as the law of inertia, can be stated thus: A body removed sufficiently far from other bodies continues in a state of rest or of uniform motion in a straight line. This law not only says something about the motion of the bodies, but it also indicates the reference-bodies or systems of coordinates, permissible in mechanics, which can be used in mechanical description. The visible fixed stars are bodies for which the law of inertia certainly holds to a high degree of approximation. Now if we use a system of co-ordinates which is rigidly attached to the earth, then, relative to this system, every fixed star describes a circle of immense radius in the course of an astronomical day, a result which is opposed to the statement of the law of inertia. So that if we adhere to this law we must refer these motions only to systems of coordinates relative to which the fixed stars do not move in a circle. A system of co-ordinates of which the state of motion is such that the law of inertia holds relative to it is called a "Galileian system of co-ordinates." The laws of the mechanics of Galilei-Newton can be regarded as valid only for a Galileian system of co-ordinates.

5. THE PRINCIPLE OF RELATIVITY (IN THE RESTRICTED SENSE)

In order to attain the greatest possible clearness, let us return to our example of the railway carriage supposed to be travelling uniformly. We call its motion a uniform translation ("uniform" because it is of constant velocity and direction, "translation" because although the carriage changes its position relative to the embankment yet it does not rotate in so doing). Let us imagine a raven flying through the air in such a manner that its motion, as observed from the embankment, is uniform and in a straight line. If we were to observe the flying raven from the moving railway carriage. we should find that the motion of the raven would be one of different velocity and direction, but that it would still be uniform and in a straight line. Expressed in an abstract manner we may say : If a mass m is moving uniformly in a straight line with respect to a co-ordinate system K, then it will also be moving uniformly and in a straight line relative to a second co-ordinate system K' provided that the latter is executing a uniform translatory motion with respect to K. In accordance with the discussion contained in the preceding section, it follows that:

If K is a Galileian co-ordinate system. then every other co-ordinate system K' is a Galileian one, when, in relation to K, it is in a condition of uniform motion of translation. Relative to K' the mechanical laws of Galilei-Newton hold good exactly as they do with respect to K.

We advance a step farther in our generalisation when we express the tenet thus: If, relative to K, K' is a uniformly moving co-ordinate system devoid of rotation, then natural phenomena run their course with respect to K' according to exactly the same general laws as with respect to K. This statement is called the principle of relativity (in the restricted sense).

As long as one was convinced that all natural phenomena were capableof representation with the help of classical mechanics, there was no need to doubt the validity of this principle of relativity. But in view of the more recent development of electrodynamics and optics it became more and more evident that classical mechanics affords an insufficient foundation for the physical description of all natural phenomena. At this juncture the question of the validity of the principle of relativity became ripe for discussion, and it did not appear impossible that the answer to this question might be in the negative.

Nevertheless, there are two general facts which at the outset speak very much in favour of the validity of the principle of relativity. Even though classical mechanics does not supply us with a sufficiently broad basis for the theoretical presentation of all physical phenomena, still we must grant it a considerable measure of "truth," since it supplies us with the actual motions of the heavenly bodies with a delicacy of detail little short of wonderful. The principle of relativity must therefore apply with great accuracy in the domain of mechanics. But that a principle of such broad generality should hold with such exactness in one domain of phenomena, and yet should be invalid for another, is a priori not very probable.

We now proceed to the second argument, to which, moreover, we shall return later. If the principle of relativity (in the restricted sense) does not hold, then the Galileian co-ordinate systems K, K', K", etc., which are moving uniformly relative to each other, will not be equivalent for the description of natural phenomena. In this case we should be constrained to believe that natural laws are capable of being formulated in a particularly simple manner, and of course only on condition that, from amongst all possible Galileian co-ordinate systems, we should have chosen one (K_0) of a particular state of motion as our body of reference. We should then be justified (because of its merits for the description of natural phenomena) in calling this system "absolutely at rest," and all other Galileian systems K

"in motion." If, for instance, our embankment were the system K_0 then our railway carriage would be a system K, relative to which less simple laws would hold than with respect to K_0. This diminished simplicity would be due to the fact that the carriage K would be in motion (i.e."really")with respect to K_0. In the general laws of nature which have been formulated with reference to K, the magnitude and direction of the velocity of the carriage would necessarily play a part. We should expect, for instance, that the note emitted by an organpipe placed with its axis parallel to the direction of travel would be different from that emitted if the axis of the pipe were placed perpendicular to this direction.

Now in virtue of its motion in an orbit round the sun, our earth is comparable with a railway carriage travelling with a velocity of about 30 kilometres per second. If the principle of relativity were not valid we should therefore expect that the direction of motion of the earth at any moment would enter into the laws of nature, and also that physical systems in their behaviour would be dependent on the orientation in space with respect to the earth. For owing to the alteration in direction of the velocity of revolution of the earth in the course of a year, the earth cannot be at rest relative to the hypothetical system K_0 throughout the whole year. However, the most careful observations have never revealed such anisotropic properties in terrestrial physical space, i.e. a physical non-equivalence of different directions. This is very powerful argument in favour of the principle of relativity.

6. THE THEOREM OF THE ADDITION OF VELOCITIES EMPLOYED IN CLASSICAL MECHANICS

Let us suppose our old friend the railway carriage to be travelling along the rails with a constant velocity v, and that a man traverses the length of the carriage in the direction of travel with a velocity w. How quickly or, in other words, with what velocity W does the man advance relative to the embankment during the process? The only possible answer seems to result from the following consideration: If the man were to stand still for a second, he would advance relative to the embankment through a distance v equal numerically to the velocity of the carriage. As a consequence of his walking, however, he traverses an additional distance w relative to the carriage, and hence also relative to the embankment, in this second, the distance w being numerically equal to the velocity with which he is walking. Thus in total he covers the distance W=v+w relative to the embankment in the second considered. We shall see later that this result, which expresses the theorem of the addition of velocities employed in classical mechanics, cannot be maintained ; in other words, the law that we have just written down does not hold in reality. For the time being, however, we shall assume its correctness.

7. THE APPARENT INCOMPATIBILITY OF THE LAW OF PROPAGATION OF LIGHT WITH THE
PRINCIPLE OF RELATIVITY

There is hardly a simpler law in physics than that according to which light is propagated in empty space. Every child at school knows, or believes he knows, that this propagation takes place in straight lines with a velocity c= 300,000 km./sec. At all events we know with great exactness that this velocity is the same for all colours, because if this were not the case, the minimum of emission would not be observed simultaneously for different colours during the eclipse of a fixed star by its dark neighbour. By means of similar considerations based on observations of double stars, the Dutch astronomer De Sitter was also able to show that the velocity of propagation of light cannot depend on the velocity of motion of the body emitting the light. The assumption that this velocity of propagation is dependent on the direction "in space" is in itself improbable.

In short, let us assume that the simple law of the constancy of the velocity of light c (in vacuum) is justifiably believed by the child at school. Who would imagine that this simple law has plunged the conscientiously thoughtful physicist into the greatest intellectual difficulties? Let us consider how these difficulties arise.

Of course we must refer the process of the propagation of light (and indeed every other process) to a rigid reference-body (co-ordinate system). As such a system let us again choose our embankment. We shall imagine the air above it to have been removed. If a ray of light be sent along the embankment, we see from the above that the tip of the ray will be transmitted with the velocity c relative to the embankment. Now let us suppose that our railway carriage is again travelling along the railway lines with the velocity v, and that its direction is the same as that of the ray of light, but its velocity of course much less. Let us inquire about the velocity of propagation of the ray of light relative to the carriage. It is obvious that we can here apply the consideration of the previous section, since the ray of light plays the part of the man walking along relatively to the carriage. The velocity w of the man relative to the embankment is here replaced by the velocity of light relative to the embankment. w is the required velocity of light with respect to the carriage, and we have

$$w = c - v.$$

The velocity of propagation ot a ray of light relative to the carriage thus comes out smaller than c.

But this result comes into conflict with the principle of relativity set forth in Section V. For, like every other general law of nature, the law of the transmission of light in vacuo [in vacuum] must, according to the principle of relativity, be the same for the railway carriage as reference-body as when the rails are the body of reference. But, from our above consideration, this would appear to be impossible. If every ray of light is propagated relative to the embankment with the velocity c, then for this reason it would appear that another law of propagation of light must necessarily hold with respect to the carriage--a result contradictory to the principle of relativity.

In view of this dilemma there appears to be nothing else for it than to abandon either the principle of relativity or the simple law of the propagation of light in vacuo. Those of you who have carefully followed the preceding discussion are almost sure to expect that we should retain the principle of relativity, which appeals so convincingly to the intellect because it is so natural and simple. The law of the propagation of light in vacuo would then have to be replaced by a more complicated law conformable to the principle of relativity. The development of theoretical physics shows, however, that we cannot pursue this course. The epoch-making theoretical investigations of H. A. Lorentz on the electrodynamical and optical phenomena connected with moving bodies show that experience in this domain leads conclusively to a theory of electromagnetic phenomena, of which the law of the constancy of the velocity of light in vacuo is a necessary consequence. Prominent theoretical physicists were therefore more inclined to reject the principle of relativity, in spite of the fact that no empirical data had been found which were contradictory to this principle.

At this juncture the theory of relativity entered the arena. As a result of an analysis of the physical conceptions of time and space, it became evident that in reality there is not the least incompatibilitiy between the principle of relativity and the law of propagation of light, and that by systematically holding fast to both these laws a logically rigid theory could be arrived at. This theory has been called the special theory of relativity to distinguish it from the extended theory, with which we shall deal later. In the following pages we shall present the fundamental ideas of the special theory of relativity.

8. ON THE IDEA OF TIME IN PHYSICS

Lightning has struck the rails on our railway embankment at two places A and B far distant from each other. I make the additional assertion that these two lightning flashes occurred simultaneously. If I ask you whether there is sense in this statement, you will answer my question with a decided "Yes." But if I now approach you with the request to explain to me the sense of the statement more precisely, you find after some consideration that the answer to this question is not so easy as it appears at first sight.

After some time perhaps the following answer would occur to you: "The significance of the statement is clear in itself and needs no further explanation; of course it would require some consideration if I were to be commissioned to determine by observations whether in the actual case the two events took place simultaneously or not." I cannot be satisfied with this answer for the following reason. Supposing that as a result of ingenious considerations an able meteorologist were to discover that the lightning must always strike the places A and B simultaneously, then we should be faced with the task of testing whether or not this theoretical result is in accordance with the reality. We encounter the same difficulty with all physical statements in which the conception "simultaneous" plays a part. The concept does not exist for the physicist until he has the possibility of discovering whether or not it is fulfilled in an actual case. We thus require a definition of simultaneity such that this definition supplies us with the method by means of which, in the present case, he can decide by experiment whether or not both the lightning strokes occurred simultaneously. As long as this requirement is not satisfied, I allow myself to be deceived as a physicist (and of course the same applies if I am not a physicist), when I imagine that I am able to attach a meaning to the statement of simultaneity. (I would ask the reader not to proceed farther until he is fully convinced on this point.)

After thinking the matter over for some time you then offer the following suggestion with which to test simultaneity. By measuring along the rails, the connecting line AB should be measured up and an observer placed at the mid-point M of the distance AB. This observer should be supplied with an arrangement (e.g. two mirrors inclined at 90°) which allows him visually to observe both places A and B at the same time. If the observer perceives the two flashes of lightning at the same time, then they are simultaneous.

I am very pleased with this suggestion, but for all that I cannot regard the matter as quite settled, because I feel constrained to raise the following objection:

"Your definition would certainly be right, if only I knew that the light by means of which the observer at M perceives the lightning flashes travels along the length $A \rightarrow M$ with the same velocity as along the length $B \rightarrow M$. But an examination of this supposition would only be possible if we already had at our disposal the means of measuring time. It would thus appear as though we were moving here in a logical circle."

After further consideration you cast a somewhat disdainful glance at me--and rightly so--and you declare:

"I maintain my previous definition nevertheless, because in reality it assumes absolutely nothing about light. There is only one demand to be made of the definition of simultaneity, namely, that in every real case it must supply us with an empirical decision as to whether or not the conception that has to be defined is fulfilled. That my definition satisfies this demand is indisputable. That light requires the same time to traverse the path $A \rightarrow M$ as for the path $B \rightarrow M$ is in reality neither a supposition nor a hypothesis about the physical nature of light, but a stipulation which I can make of my own freewill in order to arrive at a definition of simultaneity."

It is clear that this definition can be used to give an exact meaning not only to two events, but to as many events as we care to choose, and independently of the positions of the scenes of the events with respect to the body of reference [7] (here the railway embankment). We are thus led also to a definition of "time" in physics. For this purpose we suppose that clocks of identical construction are placed at the points A, B and C of the railway line (co-ordinate system) and that they are set in such a manner that the positions of their pointers are simultaneously (in the above sense) the same. Under these conditions we understand by the "time" of an event the reading (position of the hands) of that one of these clocks which is in the immediate vicinity (in space) of the event. In this manner a time-value is associated with every event which is essentially capable of observation.

This stipulation contains a further physical hypothesis, the validity of which will hardly be doubted without empirical evidence to the contrary. It has been assumed that all these clocks go at the same rate if they are of identical construction. Stated more exactly: When two clocks arranged at rest in different places of a reference-body are set in such a manner that a particular position of the pointers of the one clock is simultaneous (in the above sense) with the same position, of the pointers of the other clock, then identical "settings" are always simultaneous (in the sense of the above definition).

Notes

7) We suppose further, that, when three events A, B and C occur in different places in such a manner that A is simultaneous with B and B is simultaneous with C (simultaneous in the sense of the above definition), then the criterion for the simultaneity of the pair of events A, C is also satisfied. This assumption is a physical hypothesis about the the of propagation of light: it must certainly be fulfilled if we are to maintain the law of the constancy of the velocity of light in vacuo.

9. THE RELATIVITY OF SIMULTANEITY

Up to now our considerations have been referred to a particular body of reference, which we have styled a "railway embankment." We suppose a very long train travelling along the rails with the constant velocity v and in the direction indicated in Fig 1. People travelling in this train will with a vantage view the train as a rigid reference-body (co-ordinate system); they regard all events in

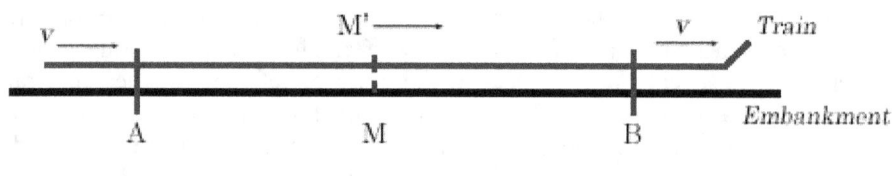

FIG. I.

reference to the train. Then every event which takes place along the line also takes place at a particular point of the train. Also the definition of simultaneity can be given relative to the train in exactly the same way as with respect to the embankment. As a natural consequence, however, the following question arises :

Are two events (e.g. the two strokes of lightning A and B) which are simultaneous with reference to the railway embankment also simultaneous relatively to the train? We shall show directly that the answer must be in the negative.

When we say that the lightning strokes A and B are simultaneous with respect to be embankment, we mean: the rays of light emitted at the places A and B, where the lightning occurs, meet each other at the mid-point M of the length A → B of the embankment. But the events A and B also correspond to positions A and B on the train. Let M' be the mid-point of the distance A→ B on the travelling train. Just when the flashes (as judged from the embankment) of lightning occur, this point M' naturally coincides with the point M but it moves towards the right in the diagram with the velocity v of the train. If an observer sitting in the position M' in the train did not possess this velocity, then he would remain permanently at M, and the light rays emitted by the flashes of lightning A and B would reach him simultaneously, i.e. they would meet just where he is situated. Now in reality (considered with reference to the railway embankment) he is hastening towards the beam of light coming from B, whilst he is riding on ahead of the beam of light coming from A. Hence the observer will see the beam of light emitted from B earlier than he will see that emitted from A. Observers who take the railway train as their reference-body must therefore come to the conclusion that the lightning flash B took place earlier than the lightning flash A. We thus arrive at the important result:

Events which are simultaneous with reference to the embankment are not simultaneous with respect to the train, and vice versa (relativity of simultaneity). Every reference-body (co-ordinate system) has its own particular time; unless we are told the reference-body to which the statement of time refers, there is no meaning in a statement of the time of an event.

Now before the advent of the theory of relativity it had always tacitly been assumed in physics that the statement of time had an absolute significance, i.e. that it is independent of the state of motion of the body of reference. But we have just seen that this assumption is incompatible with the most natural definition of simultaneity; if we discard this assumption, then the conflict between the law of the propagation of light in vacuo and the principle of relativity (developed in Section 7) disappears.

We were led to that conflict by the considerations of Section 6, which are now no longer tenable. In that section we concluded that the man in the carriage, who traverses the distance w per second relative to the carriage, traverses the same distance also with respect to the embankment in each second of time. But, according to the foregoing considerations, the time required by a particular occurrence with respect to the carriage must not be considered equal to the duration of the same occurrence as judged from the embankment (as reference-body). Hence it cannot be contended that the man in walking travels the distance w relative to the railway line in a time which is equal to one second as judged from the embankment.

Moreover, the considerations of Section 6 are based on yet a second assumption, which, in the light of a strict consideration, appears to be arbitrary, although it was always tacitly made even before the introduction of the theory of relativity.

10. ON THE RELATIVITY OF THE CONCEPTION OF DISTANCE

Let us consider two particular points on the train [8] travelling along the embankment with the velocity v, and inquire as to their distance apart. We already know that it is necessary to have a body of reference for the measurement of a distance, with respect to which body the distance can be measured up. It is the simplest plan to use the train itself as reference-body (co-ordinate system). An observer in the train measures the interval by marking off his measuring-rod in a straight line (e.g. along the floor of the carriage) as many times as is necessary to take him from the one marked point to the other. Then the number which tells us how often the rod has to be laid down is the required distance.

It is a different matter when the distance has to be judged from the railway line. Here the following method suggests itself. If we call A' and B' the two points on the train whose distance apart is required, then both of these points are moving with the velocity v along the embankment. In the first place we require to determine the points A and B of the embankment which are just being passed by the two points A' and B' at a particular time t--judged from the embankment. These points A and B of the embankment can be determined by applying the definition of time given in Section 8. The distance between these points A and B is then measured by repeated application of the measuring-rod along the embankment.

A priori it is by no means certain that this last measurement will supply us with the same result as the first. Thus the length of the train as measured from the embankment may be different from that obtained by measuring in the train itself. This circumstance leads us to a second objection which must be raised against the apparently obvious consideration of Section 6. Namely, if the man in the carriage covers the distance w in a unit of time--measured from the train,--then this distance--as measured from the embankment – is not necessarily also equal to w.

Notes

8) e.g. the middle of the first and of the hundredth carriage.

11. THE LORENTZ TRANSFORMATION

The results of the last three sections show that the apparent incompatibility of the law of propagation of light with the principle of relativity (Section 7) has been derived by means of a consideration which borrowed two unjustifiable hypotheses from classical mechanics; these are as follows:

(1) The time-interval (time) between two events is independent of the condition of motion of the body of reference.

(2) The space-interval (distance) between two points of a rigid body is independent of the condition of motion of the body of reference.

If we drop these hypotheses, then the dilemma of Section 7 disappears, because the theorem of the addition of velocities derived in Section 6 becomes invalid. The possibility presents itself that the law of the propagation of light in vacuo may be compatible with the principle of relativity, and the question arises: How have we to modify the considerations of Section 6 in order to remove the apparent disagreement between these two fundamental results of experience? This question leads to a general one. In the discussion of Section 6 we have to do with places and times relative both to the train and to the embankment. How are we to find the place and time of an event in relation to the train, when we know the place and time of the event with respect to the railway embankment? Is there a thinkable answer to this question of such a nature that the law of transmission of light in vacuo does not contradict the principle of relativity? In other words : Can we conceive of a relation between place and time of the individual events relative to both reference-bodies, such that every ray of light possesses the velocity of transmission c relative to the embankment and relative to the train? This question leads to a quite definite positive answer, and to a perfectly definite transformation law for the space-time magnitudes of an event when changing over from one body of reference to another.

Before we deal with this, we shall introduce the following incidental consideration. Up to the present we have only considered events taking place along the embankment, which had mathematically to assume the function of a straight line. In the manner indicated in Section 2 we can imagine this reference-body supplemented laterally and in a vertical direction by means of a framework of rods, so that an event which takes place anywhere can be localised with reference to this framework.

Similarly, we can imagine the train travelling with the velocity v to be continued across the whole of space, so that every event, no matter how far off it may be, could also be localised with respect to the second framework. Without committing any fundamental error, we can disregard the fact that in reality these frameworks would continually interfere with each other, owing to the impenetrability of solid bodies. In every such framework we imagine three surfaces

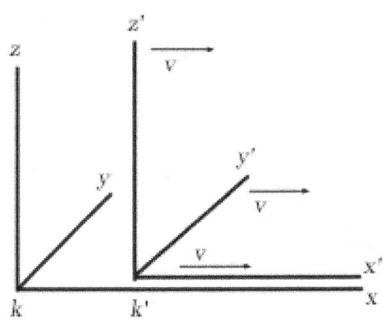

Fig. 2

243

perpendicular to each other marked out, and designated as "co-ordinate planes" ("co-ordinate system"). A co-ordinate system K then corresponds to the embankment, and a co-ordinate system K' to the train. An event, wherever it may have taken place, would be fixed in space with respect to K by the three perpendiculars x, y, z on the co-ordinate planes, and with regard to time by a time value t. Relative to K', the same event would be fixed in respect of space and time by corresponding values x', y', z', t', which of course are not identical with x, y, z, t. It has already been set forth in detail how these magnitudes are to be regarded as results of physical measurements.

Obviously our problem can be exactly formulated in the following manner. What are the values x', y', z', t', of an event with respect to K', when the magnitudes x, y, z, t, of the same event with respect to K are given ? The relations must be so chosen that the law of the transmission of light in vacuo is satisfied for one and the same ray of light (and of course for every ray) with respect to K and K'. For the relative orientation in space of the co-ordinate systems indicated in the diagram (Fig. 2), this problem is solved by means of the equations :

$$x' = \frac{x - vt}{\sqrt{1 - \frac{v^2}{c^2}}}$$

$$y' = y$$
$$z' = z$$

$$t' = \frac{t - \frac{vx}{c^2}}{\sqrt{1 - \frac{v^2}{c^2}}}$$

This system of equations is known as the "Lorentz transformation." [9]

If in place of the law of transmission of light we had taken as our basis the tacit assumptions of the older mechanics as to the absolute character of times and lengths, then instead of the above we should have obtained the following equations:

$$x' = x - vt$$
$$y' = y$$
$$z' = z$$
$$t' = t$$

This system of equations is often termed the "Galilei transformation." The Galilei transformation can be obtained from the Lorentz transformation by substituting an infinitely large value for the velocity of light c in the latter transformation.

Aided by the following illustration, we can readily see that, in accordance with the Lorentz transformation, the law of the transmission of light in vacuo is satisfied both for the reference-body K and for the reference-body K'. A light-signal is sent along the positive x-axis, and this light-stimulus advances in accordance with the equation

$$x = ct,$$

i.e. with the velocity c. According to the equations of the Lorentz transformation, this simple relation between x and t involves a relation between x' and t'. In point of fact, if we substitute for x the value ct in the first and fourth equations of the Lorentz transformation, we obtain:

$$x' = \frac{(c - v)t}{\sqrt{1 - \frac{v^2}{c^2}}}$$

$$t' = \frac{\left(1 - \frac{v}{c}\right)t}{\sqrt{1 - \frac{v^2}{c^2}}}$$

from which, by division, the expression

$$x' = ct'$$

immediately follows. If referred to the system K', the propagation of light takes place according to this equation. We thus see that the velocity of transmission relative to the reference-body K' is also equal to c. The same result is obtained for rays of light advancing in any other direction whatsoever. Of cause this is not surprising, since the equations of the Lorentz transformation were derived conformably to this point of view.

Notes

9) A simple derivation of the Lorentz transformation is given in Appendix I.

12. THE BEHAVIOUR OF MEASURING-RODS AND CLOCKS IN MOTION

Place a metre-rod in the x'-axis of K' in such a manner that one end (the beginning) coincides with the point x'=0 whilst the other end (the end of the rod) coincides with the point x'=1. What is the length of the metre-rod relatively to the system K? In order to learn this, we need only ask where the beginning of the rod and the end of the rod lie with respect to K at a particular time t of the system K. By means of the first equation of the Lorentz transformation the values of these two points at the time t = 0 can be shown to be

$$x_{(begining\ of\ rod)} = 0 \cdot \sqrt{1 - \frac{v^2}{c^2}}$$

$$x_{(end\ of\ rod)} = 1 \cdot \sqrt{1 - \frac{v^2}{c^2}}$$

the distance between the points $\sqrt{1 - v^2/c^2}$ being

But the metre-rod is moving with the velocity v relative to K. It therefore follows that the length of a rigid metre-rod moving in the direction of its length with a velocity v is $\sqrt{1 - v^2/c^2}$ of a metre.

The rigid rod is thus shorter when in motion than when at rest, and the more quickly it is moving, the shorter is the rod. For the velocity v=c we should have $\sqrt{1 - v^2/c^2} = 0$,

and for still greater velocities the square-root becomes imaginary. From this we conclude that in the theory of relativity the velocity c plays the part of a limiting velocity, which can neither be reached nor exceeded by any real body.

Of course this feature of the velocity c as a limiting velocity also clearly follows from the equations of the Lorentz transformation, for these became meaningless if we choose values of v greater than c.

If, on the contrary, we had considered a metre-rod at rest in the x-axis with respect to K, then we should have found that the length of the rod as judged from K' would have been $\sqrt{1 - v^2/c^2}$; this is quite in accordance with the principle of relativity which forms the basis of our considerations.

A Priori it is quite clear that we must be able to learn something about the physical behaviour of measuring-rods and clocks from the equations of transformation, for the magnitudes z, y, x, t, are nothing more nor less than the results of measurements obtainable by means of measuring-rods and clocks. If we had based our considerations on the Galileian transformation we should not have obtained a contraction of the rod as a consequence of its motion.

Let us now consider a seconds-clock which is permanently situated at the origin (x'=0) of K'. t'=0 and t'=1 are two successive ticks of this clock. The first and fourth equations of the Lorentz transformation give for these two ticks :

$$t = 0$$

and

$$t' = \frac{1}{\sqrt{1 - \frac{v^2}{c^2}}}$$

As judged from K, the clock is moving with the velocity v; as judged from this reference-body, the time which elapses between two strokes of the clock is not one second, but

$$\frac{1}{\sqrt{1 - \frac{v^2}{c^2}}}$$

seconds, i.e. a somewhat larger time. As a consequence of its motion the clock goes more slowly than when at rest. Here also the velocity c plays the part of an unattainable limiting velocity.

13. THEOREM OF THE ADDITION OF VELOCITIES.

THE EXPERIMENT OF FIZEAU

Now in practice we can move clocks and measuring-rods only with velocities that are small compared with the velocity of light; hence we shall hardly be able to compare the results of the previous section directly with the reality. But, on the other hand, these results must strike you as being very singular, and for that reason I shall now draw another conclusion from the theory, one which can easily be derived from the foregoing considerations, and which has been most elegantly confirmed by experiment.

In Section 6 we derived the theorem of the addition of velocities in one direction in the form which also results from the hypotheses of classical mechanics - This theorem can also be deduced readily from the Galilei transformation (Section 11). In place of the man walking inside the carriage, we introduce a point moving relatively to the co-ordinate system K' in accordance with the equation

$$x' = wt'$$

By means of the first and fourth equations of the Galilei transformation we can express x' and t' in terms of x and t, and we then obtain

$$x = (v + w)t$$

This equation expresses nothing else than the law of motion of the point with reference to the system K (of the man with reference to the embankment). We denote this velocity by the symbol W, and we then obtain, as in Section 6,

$$W = v + w \qquad A)$$

But we can carry out this consideration just as well on the basis of the theory of relativity. In the equation

$$x' = wt' \qquad B)$$

we must then express x'and t' in terms of x and t, making use of the first and fourth equations of the Lorentz transformation. Instead of the equation (A) we then obtain the equation

$$W = \frac{v + w}{1 + \frac{vw}{c^2}}$$

which corresponds to the theorem of addition for velocities in one direction according to the theory of relativity. The question now arises as to which of these two theorems is the better in accord with experience. On this point we are enlightened by a most important experiment which the brilliant physicist Fizeau performed more than half a century ago, and which has been repeated since then by some of the best experimental physicists, so that there can be no doubt about its result. The experiment is concerned with the following question. Light travels in a motionless liquid with a particular velocity w. How quickly does it travel in the direction of the arrow in the tube T

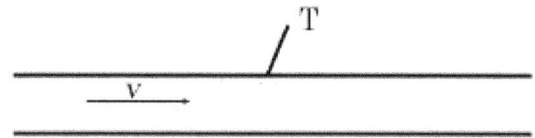

Fig. 3

when the liquid above mentioned is flowing through the tube with a velocity v ?

In accordance with the principle of relativity we shall certainly have to take for granted that the propagation of light always takes place with the same velocity w with respect to the liquid, whether the latter is in motion with reference to other bodies or not. The velocity of light relative to the liquid and the velocity of the latter relative to the tube are thus known, and we require the velocity of light relative to the tube.

It is clear that we have the problem of Section 6 again before us. The tube plays the part of the railway embankment or of the co-ordinate system K, the liquid plays the part of the carriage or of the co-ordinate system K', and finally, the light plays the part of the man walking along the carriage, or of the moving point in the present section. If we denote the velocity of the light relative to the tube by W, then this is given by the equation (A) or (B), according as the Galilei transformation or the Lorentz transformation corresponds to the facts. Experiment [10] decides in favour of equation (B) derived from the theory of relativity, and the agreement is, indeed, very exact. According to recent and most excellent measurements by Zeeman, the influence of the velocity of flow v on the propagation of light is represented by formula (B) to within one per cent.

Nevertheless we must now draw attention to the fact that a theory of this phenomenon was given by H. A. Lorentz long before the statement of the theory of relativity. This theory was of a purely electrodynamical nature, and was obtained by the use of particular hypotheses as to the electromagnetic structure of matter. This circumstance, however, does not in the least diminish the conclusiveness of the experiment as a crucial test in favour of the theory of relativity, for the electrodynamics of Maxwell-Lorentz, on which the original theory was based, in no way opposes the theory of relativity. Rather has the latter been developed trom electrodynamics as an astoundingly simple combination and generalisation of the hypotheses, formerly independent of each other, on which electrodynamics was built.

Notes

10) Fizeau found $W = w + v \left(1 - \frac{1}{n^2} \right)$, where $n = \frac{c}{w}$
is the index of refraction of the liquid. On the other hand, owing to the smallness of $\frac{vw}{c^2}$ as compared with 1,

we can replace (B) in the first place by $W = (w + v) \left(1 - \frac{vw}{c^2} \right)$, or to the same order of approximation by $w + v \left(1 - \frac{1}{n^2} \right)$, which agrees with Fizeau's result.

14. THE HEURISTIC VALUE OF THE THEORY OF RELATIVITY

Our train of thought in the foregoing pages can be epitomised in the following manner. Experience has led to the conviction that, on the one hand, the principle of relativity holds true and that on the other hand the velocity of transmission of light in vacuo has to be considered equal to a constant c. By uniting these two postulates we obtained the law of transformation for the rectangular co-ordinates x, y, z and the time t of the events which constitute the processes of nature. In this connection we did not obtain the Galilei transformation, but, differing from classical mechanics, the Lorentz transformation.

The law of transmission of light, the acceptance of which is justified by our actual knowledge, played an important part in this process of thought. Once in possession of the Lorentz transformation, however, we can combine this with the principle of relativity, and sum up the theory thus:

Every general law of nature must be so constituted that it is transformed into a law of exactly the same form when, instead of the space-time variables x, y, z, t of the original coordinate system K, we introduce new space-time variables x', y', z', t' of a co-ordinate system K'. In this connection the relation between the ordinary and the accented magnitudes is given by the Lorentz transformation. Or in brief : General laws of nature are co-variant with respect to Lorentz transformations.

This is a definite mathematical condition that the theory of relativity demands of a natural law, and in virtue of this, the theory becomes a valuable heuristic aid in the search for general laws of nature. If a general law of nature were to be found which did not satisfy this condition, then at least one of the two fundamental assumptions of the theory would have been disproved. Let us now examine what general results the latter theory has hitherto evinced.

15. GENERAL RESULTS OF THE THEORY

It is clear from our previous considerations that the (special) theory of relativity has grown out of electrodynamics and optics. In these fields it has not appreciably altered the predictions of theory, but it has considerably simplified the theoretical structure, i.e. the derivation of laws, and--what is incomparably more important – it has considerably reduced the number of independent hypothese forming the basis of theory. The special theory of relativity has rendered the Maxwell-Lorentz theory so plausible, that the latter would have been generally accepted by physicists even if experiment had decided less unequivocally in its favour.

Classical mechanics required to be modified before it could come into line with the demands of the special theory of relativity. For the main part, however, this modification affects only the laws for rapid motions, in which the velocities of matter v are not very small as compared with the velocity of light. We have experience of such rapid motions only in the case of electrons and ions; for other motions the variations from the laws of classical mechanics are too small to make themselves evident in practice. We shall not consider the motion of stars until we come to speak of the general theory of relativity. In accordance with the theory of relativity the kinetic energy of a material point of mass m is no longer given by the well-known expression

$$m \frac{v^2}{2}$$

but by the expression

$$\frac{mc^2}{\sqrt{1 - \frac{v^2}{c^2}}}$$

This expression approaches infinity as the velocity v approaches the velocity of light c. The velocity must therefore always remain less than c, however great may be the energies used to produce the acceleration. If we develop the expression for the kinetic energy in the form of a series, we obtain

$$mc^2 + m \frac{v^2}{2} + \frac{3}{8} m \frac{v^4}{c^2} + \dots$$

When $\frac{v^2}{c^2}$ is small compared with unity, the third of these terms is always small in comparison with the second,

which last is alone considered in classical mechanics. The first term mc^2 does not contain the velocity, and requires no consideration if we are only dealing with the question as to how the energy of a point-mass; depends on the velocity. We shall speak of its essential significance later.

The most important result of a general character to which the special theory of relativity has led is concerned with the conception of mass. Before the advent of relativity, physics recognised two conservation laws of fundamental importance, namely, the law of the conservation of energy and the law of the conservation of mass these two fundamental laws appeared to be quite independent of each other. By means of the theory of relativity they have been united into one law. We shall now briefly consider how this unification came about, and what meaning is to be attached to it.

The principle of relativity requires that the law of the conservation of energy should hold not only with reference to a co-ordinate system K, but also with respect to every co-ordinate system K' which is in a state of uniform motion of translation relative to K, or, briefly, relative to every "Galileian" system of co-ordinates. In contrast to classical mechanics; the Lorentz transformation is the deciding factor in the transition from one such system to another.

By means of comparatively simple considerations we are led to draw the following conclusion from these premises, in conjunction with the fundamental equations of the electrodynamics of Maxwell: A body moving with the velocity v, which absorbs [11] an amount of energy E_0 in the form of radiation without suffering an alteration in velocity in the process, has, as a consequence, its energy increased by an amount

$$\frac{E_0}{\sqrt{1-\frac{v^2}{c^2}}}$$

In consideration of the expression given above for the kinetic energy of the body, the required energy of the body comes out to be

$$\frac{\left(m+\frac{E_0}{c^2}\right)c^2}{\sqrt{1-\frac{v^2}{c^2}}}$$

Thus the body has the same energy as a body of mass

$$\left(m+\frac{E_0}{c^2}\right)$$

moving with the velocity v. Hence we can say: If a body takes up an amount of energy E_0, then its inertial mass increases by an amount

$$\frac{E_0}{c^2}$$

the inertial mass of a body is not a constant but varies according to the change in the energy of the body. The inertial mass of a system of bodies can even be regarded as a measure of its energy. The law of the conservation of the mass of a system becomes identical with the law of the conservation of energy, and is only valid provided that the system neither takes up nor sends out energy. Writing the expression for the energy in the form

$$\frac{mc^2+E_0}{\sqrt{1-\frac{v^2}{c^2}}}$$

we see that the term mc^2, which has hitherto attracted our attention, is nothing else than the energy possessed by the body [12] before it absorbed the energy E_0.

A direct comparison of this relation with experiment is not possible at the present time (1920; see [Note], p. 48), owing to the fact that the changes in energy E_0 to which we can Subject a system are not large enough to make themselves perceptible as a change in the inertial mass of the system.

$$\frac{E_0}{c^2}$$

is too small in comparison with the mass m, which was present before the alteration of the energy. It is owing to this circumstance that classical mechanics was able to establish successfully the conservation of mass as a law of independent validity.

Let me add a final remark of a fundamental nature. The success of the Faraday-Maxwell interpretation of electromagnetic action at a distance resulted in physicists becoming convinced that there are no such things as instantaneous actions at a distance (not involving an intermediary medium) of the type of Newton's law of gravitation.

According to the theory of relativity, action at a distance with the velocity of light always takes the place of instantaneous action at a distance or of action at a distance with an infinite velocity of transmission. This is connected with the fact that the velocity c plays a fundamental role in this theory. In Part II we shall see in what way this result becomes modified in the general theory of relativity.

Notes

11) E_0 is the energy taken up, as judged from a co-ordinate system moving with the body.

12) As judged from a co-ordinate system moving with the body.

[Note] The equation $E = mc^2$ has been thoroughly proved time and again since this time.

16. EXPERIENCE AND THE SPECIAL THEORY OF RELATIVITY

To what extent is the special theory of relativity supported by experience? This question is not easily answered for the reason already mentioned in connection with the fundamental experiment of Fizeau. The special theory of relativity has crystallised out from the Maxwell-Lorentz theory of electromagnetic phenomena. Thus all facts of experience which support the electromagnetic theory also support the theory of relativity. As being of particular importance, I mention here the fact that the theory of relativity enables us to predict the effects produced on the light reaching us from the fixed stars. These results are obtained in an exceedingly simple manner, and the effects indicated, which are due to the relative motion of the earth with reference to those fixed stars are found to be in accord with experience. We refer to the yearly movement of the apparent position of the fixed stars resulting from the motion of the earth round the sun (aberration), and to the influence of the radial components of the relative motions of the fixed stars with respect to the earth on the colour of the light reaching us from them. The latter effect manifests itself in a slight displacement of the spectral lines of the light transmitted to us from a fixed star, as compared with the position of the same spectral lines when they are produced by a terrestrial source of light (Doppler principle). The experimental arguments in favour of the Maxwell-Lorentz theory, which are at the same time arguments in favour of the theory of relativity, are too numerous to be set forth here. In reality they limit the theoretical possibilities to such an extent, that no other theory than that of Maxwell and Lorentz has been able to hold its own when tested by experience.

But there are two classes of experimental facts hitherto obtained which can be represented in the Maxwell-Lorentz theory only by the introduction of an auxiliary hypothesis, which in itself--i.e. without making use of the theory of relativity--appears extraneous.

It is known that cathode rays and the so-called β-rays emitted by radioactive substances consist of negatively electrified particles (electrons) of very small inertia and large velocity. By examining the deflection of these rays under the influence of electric and magnetic fields, we can study the law of motion of these particles very exactly.

In the theoretical treatment of these electrons, we are faced with the difficulty that electrodynamic theory of itself is unable to give an account of their nature. For since electrical masses of one sign repel each other, the negative electrical masses constituting the electron would necessarily be scattered under the influence of their mutual repulsions, unless there are forces of another kind operating between them, the nature of which has hitherto remained obscure to us. [13] If we now assume that the relative distances between the electrical masses constituting the electron remain unchanged during the motion of the electron (rigid connection in the sense of classical mechanics), we arrive at a law of motion of the electron which does not agree with experience. Guided by purely formal points of view, H. A. Lorentz was the first to introduce the hypothesis that the form of the electron experiences a contraction in the direction of motion in consequence of that motion. the contracted length being proportional to the expression

$$\sqrt{1 - v^2/c^2}$$

This, hypothesis, which is not justifiable by any electrodynamical facts, supplies us then with that particular law of motion which has been confirmed with great precision in recent years.

The theory of relativity leads to the same law of motion, without requiring any special hypothesis whatsoever as to the structure and the behaviour of the electron. We arrived at a similar conclusion in Section 13 in connection with the experiment of Fizeau, the result of which is foretold by the theory of relativity without the necessity of drawing on hypotheses as to the physical nature of the liquid.

The second class of facts to which we have alluded has reference to the question whether or not the motion of the earth in space can be made perceptible in terrestrial experiments. We have already remarked in Section 5 that all attempts of this nature led to a negative result. Before the theory of relativity was put forward, it was difficult to become reconciled to this negative result, for reasons now to be discussed. The inherited prejudices about time and space did not allow any doubt to arise as to the prime importance of the Galileian transformation for changing over from one body of reference to another. Now assuming that the Maxwell-Lorentz equations hold for a reference-body K, we then find that they do not hold for a reference-body K' moving uniformly with respect to K, if we assume that the relations of the Galileian transformation exist between the co-ordinates of K and K'. It thus appears that, of all Galileian co-ordinate systems, one (K) corresponding to a particular state of motion is physically unique. This result was interpreted physically by regarding K as at rest with respect to a hypothetical æther of space. On the other hand, all coordinate systems K' moving relatively to K were to be regarded as in motion with respect to the æther. To this motion of K' against the æther ("æther-drift " relative to K') were attributed the more complicated laws which were supposed to hold relative to K'. Strictly speaking, such an æther-drift ought also to be assumed relative to the earth, and for a long time the efforts of physicists were devoted to attempts to detect the existence of an æther-drift at the earth's surface.

In one of the most notable of these attempts Michelson devised a method which appears as though it must be decisive. Imagine two mirrors so arranged on a rigid body that the reflecting surfaces face each other. A ray of light requires a perfectly definite time T to pass from one mirror to the other and back again, if the whole system be at rest with respect to the æther. It is found by calculation, however, that a slightly different time T' is required for this process, if the body, together with the mirrors, be moving relatively to the æther. And yet another point: it is shown by calculation that for a given velocity v with reference to the æther, this time T' is different when the body is moving perpendicularly to the planes of the mirrors from that resulting when the motion is parallel to these planes. Although the estimated difference between these two times is exceedingly small, Michelson and Morley performed an experiment involving interference in which this difference should have been clearly detectable. But the experiment gave a negative result – a fact very perplexing to physicists. Lorentz and

FitzGerald rescued the theory from this difficulty by assuming that the motion of the body relative to the æther produces a contraction of the body in the direction of motion, the amount of contraction being just sufficient to compensate for the differeace in time mentioned above. Comparison with the discussion in Section 11 shows that also from the standpoint of the theory of relativity this solution of the difficulty was the right one. But on the basis of the theory of relativity the method of interpretation is incomparably more satisfactory. According to this theory there is no such thing as a "specially favoured" (unique) co-ordinate system to occasion the introduction of the æther-idea, and hence there can be no æther-drift, nor any experiment with which to demonstrate it. Here the contraction of moving bodies follows from the two fundamental principles of the theory, without the introduction of particular hypotheses ; and as the prime factor involved in this contraction we find, not the motion in itself, to which we cannot attach any meaning, but the motion with respect to the body of reference chosen in the particular case in point. Thus for a co-ordinate system moving with the earth the mirror system of Michelson and Morley is not shortened, but it is shortened for a co-ordinate system which is at rest relatively to the sun.

Notes

13) The general theory of relativity renders it likely that the electrical masses of an electron are held together by gravitational forces.

17. MINKOWSKI'S FOUR-DIMENSIONAL SPACE

The non-mathematician is seized by a mysterious shuddering when he hears of "four-dimensional" things, by a feeling not unlike that awakened by thoughts of the occult. And yet there is no more common-place statement than that the world in which we live is a four-dimensional space-time continuum.

Space is a three-dimensional continuum. By this we mean that it is possible to describe the position of a point (at rest) by means of three numbers (co-ordinates) x, y, z, and that there is an indefinite number of points in the neighbourhood of this one, the position of which can be described by co-ordinates such as x_1, y_1, z_1, which may be as near as we choose to the respective values of the co-ordinates x, y, z, of the first point. In virtue of the latter property we speak of a "continuum," and owing to the fact that there are three co-ordinates we speak of it as being "three-dimensional."

Similarly, the world of physical phenomena which was briefly called "world" by Minkowski is naturally four dimensional in the space-time sense. For it is composed of individual events, each of which is described by four numbers, namely, three space co-ordinates x, y, z, and a time co-ordinate, the time value t. The "world" is in this sense also a continuum; for to every event there are as many "neighbouring" events (realised or at least thinkable) as we care to choose, the co-ordinates x_1, y_1, z_1, t_1 of which differ by an indefinitely small amount from those of the event x, y, z, t originally considered. That we have not been accustomed to regard the world in this sense as a four-dimensional continuum is due to the fact that in physics, before the advent of the theory of relativity, time played a different and more independent role, as compared with the space coordinates. It is for this reason that we have been in the habit of treating time as an independent continuum. As a matter of fact, according to classical mechanics, time is absolute, i.e. it is independent of the position and the condition of motion of the system of co-ordinates. We see this expressed in the last equation of the Galileian transformation (t' = t).

The four-dimensional mode of consideration of the "world" is natural on the theory of relativity, since according to this theory time is robbed of its independence. This is shown by the fourth equation of the Lorentz transformation:

$$t' = \frac{t - \frac{vx}{c^2}}{\sqrt{1 - \frac{v^2}{c^2}}}$$

Moreover, according to this equation the time difference $\Delta t'$ of two events with respect to K' does not in general vanish, even when the time difference $\Delta t'$ of the same events with reference to K vanishes. Pure "space-distance" of two events with respect to K results in "time-distance " of the same events with respect to K. But the discovery of Minkowski, which was of importance for the formal development of the theory of relativity, does not lie here. It is to be found rather in the fact of his recognition that the four-dimensional space-time continuum of the theory of relativity, in its most essential formal properties, shows a pronounced relationship to the three-dimensional continuum of Euclidean geometrical space. [14] In order to give due prominence to this relationship, however, we must replace the usual time co-ordinate t by an imaginary magnitude $\sqrt{-1} \cdot ct$ proportional to it. Under these conditions, the natural laws satisfying the demands of the (special) theory of relativity assume mathematical forms, in which the time co-ordinate plays exactly the same role as the three space co-ordinates. Formally, these four co-ordinates correspond exactly to the three space co-ordinates in Euclidean geometry. It must be clear even to the non-mathematician that, as a consequence of this purely formal addition to our knowledge, the theory perforce gained clearness in no mean measure.

These inadequate remarks can give the reader only a vague notion of the important idea contributed by Minkowski. Without it the general theory of relativity, of which the fundamental ideas are developed in the following pages, would perhaps have got no farther than its long clothes. Minkowski's work is doubtless difficult of access to anyone inexperienced in mathematics, but since it is not necessary to have a very exact grasp of this work in order to understand the fundamental ideas of either the special or the general theory of relativity, I shall leave it here at present, and revert to it only towards the end of Part 2.

Notes

14) Cf. the somewhat more detailed discussion in Appendix II.

PART II : THE GENERAL THEORY OF RELATIVITY

18. SPECIAL AND GENERAL PRINCIPLE OF RELATIVITY

The basal principle, which was the pivot of all our previous considerations, was the special principle of relativity, i.e. the principle of the physical relativity of all uniform motion. Let as once more analyse its meaning carefully.

It was at all times clear that, from the point of view of the idea it conveys to us, every motion must be considered only as a relative motion. Returning to the illustration we have frequently used of the embankment and the railway carriage, we can express the fact of the motion here taking place in the following two forms, both of which are equally justifiable :

(a) The carriage is in motion relative to the embankment,
(b) The embankment is in motion relative to the carriage.

In (a) the embankment, in (b) the carriage, serves as the body of reference in our statement of the motion taking place. If it is simply a question of detecting or of describing the motion involved, it is in principle immaterial to what reference-body we refer the motion. As already mentioned, this is self-evident, but it must not be confused with the much more comprehensive statement called "the principle of relativity," which we have taken as the basis of our investigations.

The principle we have made use of not only maintains that we may equally well choose the carriage or the embankment as our reference-body for the description of any event (for this, too, is self-evident). Our principle rather asserts what follows : If we formulate the general laws of nature as they are obtained from experience, by making use of

(a) the embankment as reference-body,
(b) the railway carriage as reference-body,

then these general laws of nature (e.g. the laws of mechanics or the law of the propagation of light in vacuo) have exactly the same form in both cases. This can also be expressed as follows : For the physical description of natural processes, neither of the reference bodies K, K' is unique (lit. "specially marked out") as compared with the other. Unlike the first, this latter statement need not of necessity hold a priori; it is not contained in the conceptions of "motion" and "reference-body" and derivable from them; only experience can decide as to its correctness or incorrectness.

Up to the present, however, we have by no means maintained the equivalence of all bodies of reference K in connection with the formulation of natural laws. Our course was more on the following Iines. In the first place, we started out from the assumption that there exists a reference-body K, whose condition of motion is such that the Galileian law holds with respect to it : A particle left to itself and sufficiently far removed from all other particles moves uniformly in a straight line. With reference to K (Galileian reference-body) the laws of nature were to be as simple as possible. But in addition to K, all bodies of reference K' should be given preference in this sense, and they should be exactly equivalent to K for the formulation of natural laws, provided that they are in a state of uniform rectilinear and non-rotary motion with respect to K ; all these bodies of reference are to be regarded as Galileian reference-bodies. The validity of the principle of relativity was assumed only for these reference-bodies, but not for others (e.g. those possessing motion of a different kind). In this sense we speak of the special principle of relativity, or special theory of relativity.

In contrast to this we wish to understand by the "general principle of relativity" the following statement : All bodies of reference K, K', etc., are equivalent for the description of natural phenomena (formulation of the general laws of nature), whatever may be their state of motion. But before proceeding farther, it ought to be pointed out that this formulation must be replaced later by a more abstract one, for reasons which will become evident at a later stage.

Since the introduction of the special principle of relativity has been justified, every intellect which strives after generalisation must feel the temptation to venture the step towards the general principle of relativity. But a simple and apparently quite reliable consideration seems to suggest that, for the present at any rate, there is little hope of success in such an attempt; Let us imagine ourselves transferred to our old friend the railway carriage, which is travelling at a uniform rate. As long as it is moving uniformly, the occupant of the carriage is not sensible of its motion, and it is for this reason that he can without reluctance interpret the facts of the case as indicating that the carriage is at rest, but the embankment in motion. Moreover, according to the special principle of relativity, this interpretation is quite justified also from a physical point of view. If the motion of the carriage is now changed into a non-uniform motion, as for instance by a powerful application of the brakes, then the occupant of the carriage experiences a correspondingly powerful jerk forwards. The retarded motion is manifested in the mechanical behaviour of bodies relative to the person in the railway carriage. The mechanical behaviour is different from that of the case previously considered, and for this reason it would appear to be impossible that the same mechanical laws hold relatively to the non-uniformly moving carriage, as hold with reference to the carriage when at rest or in uniform motion. At all events it is clear that the Galileian law does not hold with respect to the non-uniformly moving carriage. Because of this, we feel compelled at the present juncture to grant a kind of absolute physical reality to non-uniform motion, in opposition to the general principle of relativity. But in what follows we shall soon see that this conclusion cannot be maintained.

19. THE GRAVITATIONAL FIELD

"If we pick up a stone and then let it go, why does it fall to the ground ?" The usual answer to this question is: "Because it is attracted by the earth." Modern physics formulates the answer rather differently for the following reason. As a result of the more careful study of electromagnetic phenomena, we have come to regard action at a distance as a process impossible without the intervention of some intermediary medium. If, for instance, a magnet attracts a piece of iron, we cannot be content to regard this as meaning that the magnet acts directly on the iron through the intermediate empty space, but we are constrained to imagine--after the manner of Faraday--that the magnet always calls into being something physically real in the space around it, that something being what we call a "magnetic field." In its turn this magnetic field operates on the piece of iron, so that the latter strives to move towards the magnet. We shall not discuss here the justification for this incidental conception, which is indeed a somewhat arbitrary one. We shall only mention that with its aid electromagnetic phenomena can be theoretically represented much more satisfactorily than without it, and this applies particularly to the transmission of electromagnetic waves. The effects of gravitation also are regarded in an analogous manner.

The action of the earth on the stone takes place indirectly. The earth produces in its surrounding a gravitational field, which acts on the stone and produces its motion of fall. As we know from experience, the intensity of the action on a body dimishes according to a quite definite law, as we proceed farther and farther away from the earth. From our point of view this means : The law governing the properties of the gravitational field in space must be a perfectly definite one, in order correctly to represent the diminution of gravitational action with the distance from operative bodies. It is something like this: The body (e.g. the earth) produces a field in its immediate neighbourhood directly; the intensity and direction of the field at points farther removed from the body are thence determined by the law which governs the properties in space of the gravitational fields themselves.

In contrast to electric and magnetic fields, the gravitational field exhibits a most remarkable property, which is of fundamental importance for what follows. Bodies which are moving under the sole influence of a gravitational field receive an acceleration, which does not in the least depend either on the material or on the physical state of the body. For instance, a piece of lead and a piece of wood fall in exactly the same manner in a gravitational field (in vacuo), when they start off from rest or with the same initial velocity. This law, which holds most accurately, can be expressed in a different form in the light of the following consideration.

According to Newton's law of motion, we have

(Force) = (inertial mass) x (acceleration),

where the "inertial mass" is a characteristic constant of the accelerated body. If now gravitation is the cause of the acceleration, we then have

(Force) = (gravitational mass) x (intensity of the gravitational field),

where the "gravitational mass" is likewise a characteristic constant for the body. From these two relations follows:

$$(\text{acceleration}) = \frac{(\text{gravitational mass})}{(\text{inertial mass})} \times (\text{intensity of the gravitational field}).$$

If now, as we find from experience, the acceleration is to be independent of the nature and the condition of the body and always the same for a given gravitational field, then the ratio of the gravitational to the inertial mass must likewise be the same for all bodies. By a suitable choice of units we can thus make this ratio equal to unity. We then have the following law: The gravitational mass of a body is equal to its inertial law.

It is true that this important law had hitherto been recorded in mechanics, but it had not been interpreted. A satisfactory interpretation can be obtained only if we recognise the following fact : The same quality of a body manifests itself according to circumstances as "inertia" or as "weight" (lit. "heaviness"). In the following section we shall show to what extent this is actually the case, and how this question is connected with the general postulate of relativity.

20. THE EQUALITY OF INERTIAL AND GRAVITATIONAL MASS AS AN ARGUMENT FOR THE GENERAL POSTULATE OF RELATIVITY

We imagine a large portion of empty space, so far removed from stars and other appreciable masses, that we have before us approximately the conditions required by the fundamental law of Galilei. It is then possible to choose a Galileian reference-body for this part of space (world), relative to which points at rest remain at rest and points in motion continue permanently in uniform rectilinear motion. As reference-body let us imagine a spacious chest resembling a room with an observer inside who is equipped with apparatus. Gravitation naturally does not exist for this observer. He must fasten himself with strings to the floor, otherwise the slightest impact against the floor will cause him to rise slowly towards the ceiling of the room.

To the middle of the lid of the chest is fixed externally a hook with rope attached, and now a "being" (what kind of a being is immaterial to us) begins pulling at this with a constant force. The chest together with the observer then begin to move "upwards" with a uniformly accelerated motion. In course of time their velocity will reach unheard-of values--provided that we are viewing all this from another reference-body which is not being pulled with a rope.

But how does the man in the chest regard the Process? The acceleration of the chest will be transmitted to him by the reaction of the floor of the chest. He must therefore take up this pressure by means of his legs if he does not wish to be laid out full length on the floor. He is then standing in the chest in exactly the same way as anyone stands in a room of a home on our earth. If he releases a body which he previously had in his land, the accelertion of the chest will no longer be transmitted to this body, and for this reason the body will approach the floor of the chest with an accelerated relative motion. The observer will further convince himself that the acceleration of the body towards the floor of the chest is always of the same magnitude, whatever kind of body he may happen to use for the experiment.

Relying on his knowledge of the gravitational field (as it was discussed in the preceding section), the man in the chest will thus come to the conclusion that he and the chest are in a gravitational field which is constant with regard to time. Of course he will be puzzled for a moment as to why the chest does not fall in this gravitational field. just then, however, he discovers the hook in the middle of the lid of the chest and the rope which is attached to it, and he consequently comes to the conclusion that the chest is suspended at rest in the gravitational field.

Ought we to smile at the man and say that he errs in his conclusion? I do not believe we ought to if we wish to remain consistent; we must rather admit that his mode of grasping the situation violates neither reason nor known mechanical laws. Even though it is being accelerated with respect to the "Galileian space" first considered, we can nevertheless regard the chest as being at rest. We have thus good grounds for extending the principle of relativity to include bodies of reference which are accelerated with respect to each other, and as a result we have gained a powerful argument for a generalised postulate of relativity.

We must note carefully that the possibility of this mode of interpretation rests on the fundamental property of the gravitational field of giving all bodies the same acceleration, or, what comes to the same thing, on the law of the equality of inertial and gravitational mass. If this natural law did not exist, the man in the accelerated chest would not be able to interpret the behaviour of the bodies around him on the supposition of a gravitational field, and he would not be justified on the grounds of experience in supposing his reference-body to be "at rest."

Suppose that the man in the chest fixes a rope to the inner side of the lid, and that he attaches a body to the free end of the rope. The result of this will be to stretch the rope so that it will hang "vertically " downwards. If we ask for an opinion of the cause of tension in the rope, the man in the chest will say: "The suspended body experiences a downward force in the gravitational field, and this is neutralised by the tension of the rope; what determines the magnitude of the tension of the rope is the gravitational mass of the suspended body." On the other hand, an observer who is poised freely in space will interpret the condition of things thus : "The rope must perforce take part in the accelerated motion of the chest, and it transmits this motion to the body attached to it. The tension of the rope is just large enough to effect the acceleration of the body. That which determines the magnitude of the tension of the rope is the inertial mass of the body." Guided by this example, we see that our extension of the principle of relativity implies the necessity of the law of the equality of inertial and gravitational mass. Thus we have obtained a physical interpretation of this law.

From our consideration of the accelerated chest we see that a general theory of relativity must yield important results on the laws of gravitation. In point of fact, the systematic pursuit of the general idea of relativity has supplied the laws satisfied by the gravitational field. Before proceeding farther, however, I must warn the reader against a misconception suggested by these considerations. A gravitational field exists for the man in the chest, despite the fact that there was no such field for the co-ordinate system first chosen. Now we might easily suppose that the existence of a gravitational field is always only an apparent one. We might also think that, regardless of the kind of gravitational field which may be present, we could always choose another reference-body such that no gravitational field exists with reference to it. This is by no means true for all gravitational fields, but only for those of quite special form. It is, for instance, impossible to choose a body of reference such that, as judged from it, the gravitational field of the earth (in its entirety) vanishes.

We can now appreciate why that argument is not convincing, which we brought forward against the general principle of relativity at the end of Section 18. It is certainly true that the observer in the railway carriage experiences a jerk forwards as a result of the application of the brake, and that he recognises, in this the non-uniformity of motion (retardation) of the carriage. But he is compelled by nobody to refer this jerk to a "real" acceleration (retardation) of the carriage. He might also interpret his experience thus: "My body of reference (the carriage) remains permanently at rest. With reference to it, however, there exists (during the period of application of the brakes) a gravitational field which is directed forwards and which is variable with respect to time. Under the influence of this field, the embankment together with the earth moves non-uniformly in such a manner that their original velocity in the backwards direction is continuously reduced."

21. IN WHAT RESPECTS ARE THE FOUNDATIONS OF CLASSICAL MECHANICS AND OF THE SPECIAL THEORY OF RELATIVITY UNSATISFACTORY?

We have already stated several times that classical mechanics starts out from the following law: Material particles sufficiently far removed from other material particles continue to move uniformly in a straight line or continue in a state of rest. We have also repeatedly emphasised that this fundamental law can only be valid for bodies of reference K which possess certain unique states of motion, and which are in uniform translational motion relative to each other. Relative to other reference-bodies K the law is not valid. Both in classical mechanics and in the special theory of relativity we therefore differentiate between reference-bodies K relative to which the recognised "laws of nature" can be said to hold, and reference-bodies K relative to which these laws do not hold.

But no person whose mode of thought is logical can rest satisfied with this condition of things. He asks : "How does it come that certain reference-bodies (or their states of motion) are given priority over other reference-bodies (or their states of motion)? What is the reason for this Preference?" In order to show clearly what I mean by this question, I shall make use of a comparison.

I am standing in front of a gas range. Standing alongside of each other on the range are two pans so much alike that one may be mistaken for the other. Both are half full of water. I notice that steam is being emitted continuously from the one pan, but not from the other. I am surprised at this, even if I have never seen either a gas range or a pan before. But if I now notice a luminous something of bluish colour under the first pan but not under the other, I cease to be astonished, even if I have never before seen a gas flame. For I can only say that this bluish something will cause the emission of the steam, or at least possibly it may do so. If, however, I notice the bluish something in neither case, and if I observe that the one continuously emits steam whilst the other does not, then I shall remain astonished and dissatisfied until I have discovered some circumstance to which I can attribute the different behaviour of the two pans.

Analogously, I seek in vain for a real something in classical mechanics (or in the special theory of relativity) to which I can attribute the different behaviour of bodies considered with respect to the reference systems K and K'. [15] Newton saw this objection and attempted to invalidate it, but without success. But E. Mach recognised it most clearly of all, and because of this objection he claimed that mechanics must be placed on a new basis. It can only be got rid of by means of a physics which is conformable to the general principle of relativity, since the equations of such a theory hold for every body of reference, whatever may be its state of motion.

Notes

15) The objection is of importance more especially when the state of motion of the reference-body is of such a nature that it does not require any external agency for its maintenance, e.g. in the case when the reference-body is rotating uniformly.

22. A FEW INFERENCES FROM THE GENERAL PRINCIPLE OF RELATIVITY

The considerations of Section 20 show that the general principle of relativity puts us in a position to derive properties of the gravitational field in a purely theoretical manner. Let us suppose, for instance, that we know the space-time "course" for any natural process whatsoever, as regards the manner in which it takes place in the Galileian domain relative to a Galileian body of reference K. By means of purely theoretical operations (i.e. simply by calculation) we are then able to find how this known natural process appears, as seen from a reference-body K' which is accelerated relatively to K. But since a gravitational field exists with respect to this new body of reference K', our consideration also teaches us how the gravitational field influences the process studied.

For example, we learn that a body which is in a state of uniform rectilinear motion with respect to K (in accordance with the law of Galilei) is executing an accelerated and in general curvilinear motion with respect to the accelerated reference-body K' (chest). This acceleration or curvature corresponds to the influence on the moving body of the gravitational field prevailing relatively to K. It is known that a gravitational field influences the movement of bodies in this way, so that our consideration supplies us with nothing essentially new.

However, we obtain a new result of fundamental importance when we carry out the analogous consideration for a ray of light. With respect to the Galileian reference-body K, such a ray of light is transmitted rectilinearly with the velocity c. It can easily be shown that the path of the same ray of light is no longer a straight line when we consider it with reference to the accelerated chest (reference-body K'). From this we conclude, that, in general, rays of light are propagated curvilinearly in gravitational fields. In two respects this result is of great importance.

In the first place, it can be compared with the reality. Although a detailed examination of the question shows that the curvature of light rays required by the general theory of relativity is only exceedingly small for the gravitational fields at our disposal in practice, its estimated magnitude for light rays passing the sun at grazing incidence is nevertheless 1.7 seconds of arc. This ought to manifest itself in the following way. As seen from the earth, certain fixed stars appear to be in the neighbourhood of the sun, and are thus capable of observation during a total eclipse of the sun. At such times, these stars ought to appear to be displaced outwards from the sun by an amount indicated above, as compared with their apparent position in the sky when the sun is situated at another part of the heavens. The examination of the correctness or otherwise of this deduction is a problem of the greatest importance, the early solution of which is to be expected of astronomers. [16)]

In the second place our result shows that, according to the general theory of relativity, the law of the constancy of the velocity of light in vacuo, which constitutes one of the two fundamental assumptions in the special theory of relativity and to which we have already frequently referred, cannot claim any unlimited validity. A curvature of rays of light can only take place when the velocity of propagation of light varies with position. Now we might think that as a consequence of this, the special theory of relativity and with it the whole theory of relativity would be laid in the dust. But in reality this is not the case. We can only conclude that the special theory of relativity cannot claim an unlimited domain of validity ; its results hold only so long as we are able to disregard the influences of gravitational fields on the phenomena (e.g. of light).

Since it has often been contended by opponents of the theory of relativity that the special theory of relativity is overthrown by the general theory of relativity, it is perhaps advisable to make the facts of the case clearer by means of an appropriate comparison. Before the development of electrodynamics the laws of electrostatics were looked upon as the laws of electricity. At the present time we know that electric fields can be derived correctly from electrostatic considerations only for the case, which is never strictly realised, in which the electrical masses are quite at rest relatively to each other, and to the co-ordinate system. Should we be justified in saying that for this reason electrostatics is overthrown by the field-equations of Maxwell in electrodynamics? Not in the least. Electrostatics is contained in electrodynamics as a limiting case ; the laws of the latter lead directly to those of the former for the case in which the fields are invariable with regard to time. No fairer destiny could be allotted to any physical theory, than that it should of itself point out the way to the introduction of a more comprehensive theory, in which it lives on as a limiting case.

In the example of the transmission of light just dealt with, we have seen that the general theory of relativity enables us to derive theoretically the influence of a gravitational field on the course of natural processes, the laws of which are already known when a gravitational field is absent. But the most attractive problem, to the solution of which the general theory of relativity supplies the key, concerns the investigation of the laws satisfied by the gravitational field itself. Let us consider this for a moment.

We are acquainted with space-time domains which behave (approximately) in a "Galileian" fashion under suitable choice of reference-body, i.e. domains in which gravitational fields are absent. If we now refer such a domain to a reference-body K' possessing any kind of motion, then relative to K' there exists a gravitational field which is variable with respect to space and time. [17] The character of this field will of course depend on the motion chosen for K'. According to the general theory of relativity, the general law of the gravitational field must be satisfied for all gravitational fields obtainable in this way. Even though by no means all gravitationial fields can be produced in this way, yet we may entertain the hope that the general law of gravitation will be derivable from such gravitational fields of a special kind. This hope has been realised in the most beautiful manner. But between the clear vision of this goal and its actual realisation it was necessary to surmount a serious difficulty, and as this lies deep at the root of things, I dare not withhold it from the reader. We require to extend our ideas of the space-time continuum still farther.

Notes

16) By means of the star photographs of two expeditions equipped by a Joint Committee of the Royal and Royal Astronomical Societies, the existence of the deflection of light demanded by theory was first confirmed during the solar eclipse of 29th May, 1919. (Cf. Appendix III.)

17) This follows from a generalisation of the discussion in Section 20

23. BEHAVIOUR OF CLOCKS AND MEASURING-RODS ON A ROTATING BODY OF REFERENCE

Hitherto I have purposely refrained from speaking about the physical interpretation of space- and time-data in the case of the general theory of relativity. As a consequence, I am guilty of a certain slovenliness of treatment, which, as we know from the special theory of relativity, is far from being unimportant and pardonable. It is now high time that we remedy this defect; but I would mention at the outset, that this matter lays no small claims on the patience and on the power of abstraction of the reader.

We start off again from quite special cases, which we have frequently used before. Let us consider a space time domain in which no gravitational field exists relative to a reference-body K whose state of motion has been suitably chosen. K is then a Galileian reference-body as regards the domain considered, and the results of the special theory of relativity hold relative to K. Let us suppose the same domain referred to a second body of reference K', which is rotating uniformly with respect to K. In order to fix our ideas, we shall imagine K' to be in the form of a plane circular disc, which rotates uniformly in its own plane about its centre. An observer who is sitting eccentrically on the disc K' is sensible of a force which acts outwards in a radial direction, and which would be interpreted as an effect of inertia (centrifugal force) by an observer who was at rest with respect to the original reference-body K. But the observer on the disc may regard his disc as a reference-body which is "at rest" ; on the basis of the general principle of relativity he is justified in doing this. The force acting on himself, and in fact on all other bodies which are at rest relative to the disc, he regards as the effect of a gravitational field. Nevertheless, the space-distribution of this gravitational field is of a kind that would not be possible on Newton's theory of gravitation. [18] But since the observer believes in the general theory of relativity, this does not disturb him; he is quite in the right when he believes that a general law of gravitation can be formulated - a law which not only explains the motion of the stars correctly, but also the field of force experienced by himself.

The observer performs experiments on his circular disc with clocks and measuring-rods. In doing so, it is his intention to arrive at exact definitions for the signification of time- and space-data with reference to the circular disc K', these definitions being based on his observations. What will be his experience in this enterprise ?

To start with, he places one of two identically constructed clocks at the centre of the circular disc, and the other on the edge of the disc, so that they are at rest relative to it. We now ask ourselves whether both clocks go at the same rate from the standpoint of the non-rotating Galileian reference-body K. As judged from this body, the clock at the centre of the disc has no velocity, whereas the clock at the edge of the disc is in motion relative to K in consequence of the rotation. According to a result obtained in Section 12, it follows that the latter clock goes at a rate permanently slower than that of the clock at the centre of the circular disc, i.e. as observed from K. It is obvious that the same effect would be noted by an observer whom we will imagine sitting alongside his clock at the centre of the circular disc. Thus on our circular disc, or, to make the case more general, in every gravitational field, a clock will go more quickly or less quickly, according to the position in which the clock is situated (at rest). For this reason it is not possible to obtain a reasonable definition of time with the aid of clocks which are arranged at rest with respect to the body of reference. A similar difficulty presents itself when we attempt to apply our earlier definition of simultaneity in such a case, but I do not wish to go any farther into this question.

Moreover, at this stage the definition of the space co-ordinates also presents insurmountable difficulties. If the observer applies his standard measuring-rod (a rod which is short as compared with the radius of the disc) tangentially to the edge of the disc, then, as judged from the Galileian system, the length of this rod will be less than 1, since, according to Section 12, moving bodies suffer a shortening in the direction of the motion. On the other hand, the measuring-rod will not experience a shortening in length, as judged from K, if it is applied to the disc in the direction of the radius. If, then, the

observer first measures the circumference of the disc with his measuring-rod and then the diameter of the disc, on dividing the one by the other, he will not obtain as quotient the familiar number $\pi = 3.14$ . . ., but a larger number, [19] whereas of course, for a disc which is at rest with respect to K, this operation would yield π exactly. This proves that the propositions of Euclidean geometry cannot hold exactly on the rotating disc, nor in general in a gravitational field, at least if we attribute the length 1 to the rod in all positions and in every orientation. Hence the idea of a straight line also loses its meaning. We are therefore not in a position to define exactly the co-ordinates x, y, z relative to the disc by means of the method used in discussing the special theory, and as long as the co-ordinates and times of events have not been defined, we cannot assign an exact meaning to the natural laws in
which these occur.

Thus all our previous conclusions based on general relativity would appear to be called in question. In reality we must make a subtle detour in order to be able to apply the postulate of general relativity exactly. I shall prepare the reader for this in the following paragraphs.

Notes

18) The field disappears at the centre of the disc and increases proportionally to the distance from the centre as we proceed outwards.

19) Throughout this consideration we have to use the Galileian (non-rotating) system K as reference-body, since we may only assume the validity of the results of the special theory of relativity relative to K (relative to K' a gravitational field prevails).

24. EUCLIDEAN AND NON-EUCLIDEAN CONTINUUM

The surface of a marble table is spread out in front of me. I can get from any one point on this table to any other point by passing continuously from one point to a "neighbouring" one, and repeating this process a (large) number of times, or, in other words, by going from point to point without executing "jumps." I am sure the reader will appreciate with sufficient clearness what I mean here by "neighbouring" and by "jumps" (if he is not too pedantic). We express this property of the surface by describing the latter as a continuum.

Let us now imagine that a large number of little rods of equal length have been made, their lengths being small compared with the dimensions of the marble slab. When I say they are of equal length, I mean that one can be laid on any other without the ends overlapping. We next lay four of these little rods on the marble slab so that they constitute a quadrilateral figure (a square), the diagonals of which are equally long. To ensure the equality of the diagonals, we make use of a little testing-rod. To this square we add similar ones, each of which has one rod in common with the first. We proceed in like manner with each of these squares until finally the whole marble slab is laid out with squares. The arrangement is such, that each side of a square belongs to two squares and each corner to four squares.

It is a veritable wonder that we can carry out this business without getting into the greatest difficulties. We only need to think of the following. If at any moment three squares meet at a corner, then two sides of the fourth square are already laid, and, as a consequence, the arrangement of the remaining two sides of the square is already completely determined. But I am now no longer able to adjust the quadrilateral so that its diagonals may be equal. If they are equal of their own accord, then this is an especial favour of the marble slab and of the little rods, about which I can only be thankfully surprised. We must experience many such surprises if the construction is to be successful.

If everything has really gone smoothly, then I say that the points of the marble slab constitute a Euclidean continuum with respect to the little rod, which has been used as a "distance " (line-interval). By choosing one corner of a square as "origin" I can characterise every other corner of a square with reference to this origin by means of two numbers. I only need state how many rods I must pass over when, starting from the origin, I proceed towards the "right" and then "upwards," in order to arrive at the corner of the square under consideration. These two numbers are then the "Cartesian co-ordinates" of this corner with reference to the "Cartesian co-ordinate system" which is determined by the arrangement of little rods.

By making use of the following modification of this abstract experiment, we recognise that there must also be cases in which the experiment would be unsuccessful. We shall suppose that the rods "expand" by in amount proportional to the increase of temperature. We heat the central part of the marble slab, but not the periphery, in which case two of our little rods can still be brought into coincidence at every position on the table. But our construction of squares must necessarily come into disorder during the heating, because the little rods on the central region of the table expand, whereas those on the outer part do not.

With reference to our little rods--defined as unit lengths – the marble slab is no longer a Euclidean continuum, and we are also no longer in the position of defining Cartesian co-ordinates directly with their aid, since the above construction can no longer be carried out. But since there are other things which are not influenced in a similar manner to the little rods (or perhaps not at all) by the temperature of the table, it is possible quite naturally to maintain the point of view that the marble slab is a "Euclidean continuum." This can be done in a satisfactory manner by making a more subtle stipulation about the measurement or the comparison of lengths.

But if rods of every kind (i.e. of every material) were to behave in the same way as regards the influence of temperature when they are on the variably heated marble slab, and if we had no other means of detecting the effect of temperature than the geometrical behaviour of our rods in experiments analogous to the one described above, then our

best plan would be to assign the distance one to two points on the slab, provided that the ends of one of our rods could be made to coincide with these two points ; for how else should we define the distance without our proceeding being in the highest measure grossly arbitrary ? The method of Cartesian coordinates must then be discarded, and replaced by another which does not assume the validity of Euclidean geometry for rigid bodies.[20] The reader will notice that the situation depicted here corresponds to the one brought about by the general postulate of relativity (Section 23).

Notes

20) Mathematicians have been confronted with our problem in the following form. If we are given a surface (e.g. an ellipsoid) in Euclidean three-dimensional space, then there exists for this surface a two-dimensional geometry, just as much as for a plane surface. Gauss undertook the task of treating this two-dimensional geometry from first principles, without making use of the fact that the surface belongs to a Euclidean continuum of three dimensions. If we imagine constructions to be made with rigid rods in the surface (similar to that above with the marble slab), we should find that different laws hold for these from those resulting on the basis of Euclidean plane geometry. The surface is not a Euclidean continuum with respect to the rods, and we cannot define Cartesian co-ordinates in the surface. Gauss indicated the principles according to which we can treat the geometrical relationships in the surface, and thus pointed out the way to the method of Riemann of treating multi-dimensional, non-Euclidean continuum. Thus it is that mathematicians long ago solved the formal problems to which we are led by the general postulate of relativity.

25. GAUSSIAN CO-ORDINATES

According to Gauss, this combined analytical and geometrical mode of handling the problem can be arrived at in the following way. We imagine a system of arbitrary curves (see Fig. 4) drawn on the

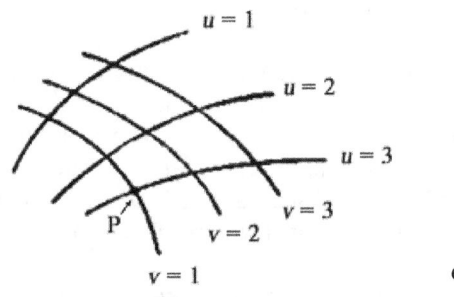

Fig. 4

273

surface of the table. These we designate as u-curves, and we indicate each of them by means of a number. The Curves u = 1, u = 2 and u = 3 are drawn in the diagram. Between the curves u = 1 and u = 2 we must imagine an infinitely large number to be drawn, all of which correspond to real numbers lying between 1 and 2. fig. 4 We have then a system of u-curves, and this "infinitely dense" system covers the whole surface of the table. These u-curves must not intersect each other, and through each point of the surface one and only one curve must pass. Thus a perfectly definite value of u belongs to every point on the surface of the marble slab. In like manner we imagine a system of v-curves drawn on the surface. These satisfy the same conditions as the u-curves, they are provided with numbers in a corresponding manner, and they may likewise be of arbitrary shape. It follows that a value of u and a value of v belong to every point on the surface of the table. We call these two numbers the co-ordinates of the surface of the table (Gaussian co-ordinates). For example, the point P in the diagram has the Gaussian co-ordinates u = 3, v = 1. Two neighbouring points P and P_1 on the surface then correspond to the co-ordinates

P: u, v

P_1:   u + du, v + dv,

where du and dv signify very small numbers. In a similar manner we may indicate the distance (line-interval) between P and P_1, as measured with a little rod, by means of the very small number ds. Then according to Gauss we have

$$ds^2 = g_{11}du^2 + 2g_{12}dudv + g_{22}dv^2$$

where g_{11}, g_{12}, g_{22}, are magnitudes which depend in a perfectly definite way on u and v. The magnitudes g_{11}, g_{12} and g_{22}, determine the behaviour of the rods relative to the u-curves and v-curves, and thus also relative to the surface of the table. For the case in which the points of the surface considered form a Euclidean continuum with reference to the measuring-rods, but only in this case, it is possible to draw the u-curves and v-curves and to attach numbers to them, in such a manner, that we simply have :

$$ds^2 = du^2 + dv^2$$

Under these conditions, the u-curves and v-curves are straight lines in the sense of Euclidean geometry, and they are perpendicular to each other. Here the Gaussian coordinates are simply Cartesian ones. It is clear that Gauss co-ordinates are nothing more than an association of two sets of numbers with the points of the surface considered, of such a nature that numerical values differing very slightly from each other are associated with neighbouring points "in space."

So far, these considerations hold for a continuum of two dimensions. But the Gaussian method can be applied also to a continuum of three, four or more dimensions.

If, for instance, a continuum of four dimensions be supposed available, we may represent it in the following way. With every point of the continuum, we associate arbitrarily four numbers, x_1, x_2, x_3, x_4, which are known as "co-ordinates." Adjacent points correspond to adjacent values of the coordinates. If a distance ds is associated with the adjacent points P and P_1, this distance being measurable and well defined from a physical point of view, then the following formula holds:

$$ds^2 = g_{11}dx_1^2 + 2g_{12}dx_1dx_2 \ldots g_{44}dx_4^2,$$

where the magnitudes g_{11}, etc., have values which vary with the position in the continuum. Only when the continuum is a Euclidean one is it possible to associate the co-ordinates $x_1 \ldots x_4$. with the points of the continuum so that we have simply

$$ds^2 = dx_1^2 + dx_2^2 + dx_3^2 + dx_4^2.$$

In this case relations hold in the four-dimensional continuum which are analogous to those holding in our three-dimensional measurements.

However, the Gauss treatment for ds^2 which we have given above is not always possible. It is only possible when sufficiently small regions of the continuum under consideration may be regarded as Euclidean continua. For example, this obviously holds in the case of the marble slab of the table and local variation of temperature. The temperature is practically constant for a small part of the slab, and thus the geometrical behaviour of the rods is almost as it ought to be according to the rules of Euclidean geometry. Hence the imperfections of the construction of squares in the previous section do not show themselves clearly until this construction is extended over a considerable portion of the surface of the table.

We can sum this up as follows: Gauss invented a method for the mathematical treatment of continua in general, in which "size-relations" ("distances" between neighbouring points) are defined. To every point of a continuum are assigned as many numbers (Gaussian coordinates) as the continuum has dimensions. This is done in such a way, that only one meaning can be attached to the assignment, and that numbers (Gaussian coordinates) which differ by an indefinitely small amount are assigned to adjacent points. The Gaussian coordinate system is a logical generalisation of the Cartesian co-ordinate system. It is also applicable to non-Euclidean continua, but only when, with respect to the defined "size" or "distance," small parts of the continuum under consideration behave more nearly like a Euclidean system, the smaller the part of the continuum under our notice.

26. THE SPACE-TIME CONTINUUM OF THE SPECIAL THEORY OF RELATIVITY CONSIDERED AS A EUCLIDEAN CONTINUUM

We are now in a position to formulate more exactly the idea of Minkowski, which was only vaguely indicated in Section 17. In accordance with the special theory of relativity, certain co-ordinate systems are given preference for the description of the four-dimensional, space-time continuum. We called these "Galileian co-ordinate systems." For these systems, the four co-ordinates x, y, z, t, which determine an event or--in other words, a point of thefour-dimensional continuum--are defined physically in a simple manner, as set forth in detail in the first part of this book. For the transition from one Galileian system to another, which is moving uniformly with reference to the first, the equations of the Lorentz transformation are valid. These last form the basis for the derivation of deductions from the special theory of relativity, and in themselves they are nothing more than the expression of the universal validity of the law of transmission of light for all Galileian systems of reference.

Minkowski found that the Lorentz transformations satisfy the following simple conditions. Let us consider two neighbouring events, the relative position of which in the four-dimensional continuum is given with respect to a Galileian reference-body K by the space co-ordinate differences dx, dy, dz and the time-difference dt. With reference to a second Galileian system we shall suppose that the corresponding differences for these two events are dx_1, dy_1, dz_1, dt_1. Then these magnitudes always fulfill the condition [21]

$$dx^2 + dy^2 + dz^2 - c^2 dt^2 = dx_1^2 + dy_1^2 + dz_1^2 - c^2 dt_1^2.$$

The validity of the Lorentz transformation follows from this condition. We can express this as follows: The magnitude

$$ds^2 = dx^2 + dy^2 + dz^2 - c^2 dt^2,$$

which belongs to two adjacent points of the four-dimensional space-time continuum, has the same value for all selected (Galileian) reference-bodies. If we replace x, y, z, $\sqrt{-1} \cdot ct$, by x_1, x_2, x_3, x_4, we also obtain the result that

$$ds^2 = dx_1^2 + dx_2^2 + dx_3^2 + dx_4^2.$$

is independent of the choice of the body of reference. We call the magnitude ds the " distance " apart of the two events or four-dimensional points.

Thus, if we choose as time-variable the imaginary variable $\sqrt{-1} \cdot ct$ instead of the real quantity t, we can regard the space-time contintium--accordance with the special theory of relativity--as a ", Euclidean" four-dimensional continuum, a result which follows from the considerations of the preceding section.

Notes

21) Cf. Appendixes I and 2. The relations which are derived there for the co-ordinates themselves are valid also for co-ordinate differences, and thus also for co-ordinate differentials (indefinitely small differences).

27. THE SPACE-TIME CONTINUUM OF THE GENERAL THEORY OF RELATIVITY IS NOT A EUCLIDEAN CONTINUUM

In the first part of this book we were able to make use of space-time co-ordinates which allowed of a simple and direct physical interpretation, and which, according to Section 26, can be regarded as four-dimensional Cartesian co-ordinates. This was possible on the basis of the law of the constancy of the velocity of light. But according to Section 21 the general theory of relativity cannot retain this law. On the contrary, we arrived at the result that according to this latter theory the velocity of light must always depend on the co-ordinates when a gravitational field is present. In connection with a specific illustration in Section 23, we found that the presence of a gravitational field invalidates the definition of the coordinates and the time, which led us to our objective in the special theory of relativity.

In view of the results of these considerations we are led to the conviction that, according to the general principle of relativity, the space-time continuum cannot be regarded as a Euclidean one, but that here we have the general case, corresponding to the marble slab with local variations of temperature, and with which we made acquaintance as an example of a two-dimensional continuum. Just as it was there impossible to construct a Cartesian co-ordinate system from equal rods, so here it is impossible to build up a system (reference-body) from rigid bodies and clocks, which shall be of such a nature that measuring-rods and clocks, arranged rigidly with respect to one another, shall indicate position and time directly. Such was the essence of the difficulty with which we were confronted in Section 23.

But the considerations of Sections 25 and 26 show us the way to surmount this difficulty. We refer the four-dimensional space-time continuum in an arbitrary manner to Gauss co-ordinates. We assign to every point of the continuum (event) four numbers, x_1, x_2, x_3, x_4 (co-ordinates), which have not the least direct physical significance, but only serve the purpose of numbering the points of the continuum in a definite but arbitrary manner. This arrangement does not even need to be of such a kind that we must regard x_1, x_2, x_3, as "space" co-ordinates and x_4, as a "time" co-ordinate.

The reader may think that such a description of the world would be quite inadequate. What does it mean to assign to an event the particular co-ordinates x_1, x_2, x_3, x_4, if in themselves these co-ordinates have no significance? More careful consideration shows, however, that this anxiety is unfounded. Let us consider, for instance, a material point with any kind of motion. If this point had only a momentary existence without duration, then it would to described in space-time by a single system of values x_1, x_2, x_3, x_4. Thus its permanent existence must be characterised by an infinitely large number of such systems of values, the co-ordinate values of which are so close together as to give continuity; corresponding to the material point, we thus have a (uni-dimensional) line in the four-dimensional continuum. In the same way, any such lines in our continuum correspond to many points in motion. The only statements having regard to these points which can claim a physical existence are in reality the statements about their encounters. In our mathematical treatment, such an encounter is expressed in the fact that the two lines which represent the motions of the points in question have a particular system of co-ordinate values, x_1, x_2, x_3, x_4, in common. After mature consideration the reader will doubtless admit that in reality such encounters constitute the only actual evidence of a time-space nature with which we meet in physical statements.

When we were describing the motion of a material point relative to a body of reference, we stated nothing more than the encounters of this point with particular points of the reference-body. We can also determine the corresponding values of the time by the observation of encounters of the body with clocks, in conjunction with the observation of the encounter of the hands of clocks with particular points on the dials. It is just the same in the case of space-measurements by means of measuring-rods, as a little consideration will show.

The following statements hold generally: Every physical description resolves itself into a number of statements, each of which refers to the space-time coincidence of two events A and B. In terms of Gaussian co-ordinates, every such statement is expressed by the agreement of their four co-ordinates x_1, x_2, x_3, x_4. Thus in reality, the description of the time-space continuum by means of Gauss co-ordinates completely replaces the description with the aid of a body of reference, without suffering from the defects of the latter mode of description; it is not tied down to the Euclidean character of the continuum which has to be represented.

28. EXACT FORMULATION OF THE GENERAL PRINCIPLE OF RELATIVITY

We are now in a position to replace the provisional formulation of the general principle of relativity given in Section 18 by an exact formulation. The form there used, "All bodies of reference K, K', etc., are equivalent for the description of natural phenomena (formulation of the general laws of nature), whatever may be their state of motion," cannot be maintained, because the use of rigid reference-bodies, in the sense of the method followed in the special theory of relativity, is in general not possible in space-time description. The Gauss co-ordinate system has to take the place of the body of reference. The following statement corresponds to the fundamental idea of the general principle of relativity: "All Gaussian co-ordinate systems are essentially equivalent for the formulation of the general laws of nature."

We can state this general principle of relativity in still another form, which renders it yet more clearly intelligible than it is when in the form of the natural extension of the special principle of relativity. According to the special theory of relativity, the equations which express the general laws of nature pass over into equations of the same form when, by making use of the Lorentz transformation, we replace the space-time variables x, y, z, t, of a (Galileian) reference-body K by the space-time variables x', y', z', t', of a new reference-body K'. According to the general theory of relativity, on the other hand, by application of arbitrary substitutions of the Gauss variables x_1, x_2, x_3, x_4, the equations must pass over into equations of the same form; for every transformation (not only the Lorentz transformation) corresponds to the transition of one Gauss co-ordinate system into another.

If we desire to adhere to our "old-time" three-dimensional view of things, then we can characterise the development which is being undergone by the fundamental idea of the general theory of relativity as follows : The special theory of relativity has reference to Galileian domains, i.e. to those in which no gravitational field exists. In this connection a Galileian reference-body serves as body of reference, i.e. a rigid body the state of motion of which is so chosen that the Galileian law of the uniform rectilinear motion of "isolated" material points holds relatively to it.

Certain considerations suggest that we should refer the same Galileian domains to non-Galileian reference-bodies also. A gravitational field of a special kind is then present with respect to these bodies (cf. Sections 20 and 23).

In gravitational fields there are no such things as rigid bodies with Euclidean properties; thus the fictitious rigid body of reference is of no avail in the general theory of relativity. The motion of clocks is also influenced by gravitational fields, and in such a way that a physical definition of time which is made directly with the aid of clocks has by no means the same degree of plausibility as in the special theory of relativity.

For this reason non-rigid reference-bodies are used, which are as a whole not only moving in any way whatsoever, but which also suffer alterations in form ad lib. during their motion. Clocks, for which the law of motion is of any kind, however irregular, serve for the definition of time. We have to imagine each of these clocks fixed at a point on the non-rigid reference-body. These clocks satisfy only the one condition, that the "readings" which are observed simultaneously on adjacent clocks (in space) differ from each other by an indefinitely small amount. This non-rigid reference-body, which might appropriately be termed a "reference-mollusc", is in the main equivalent to a Gaussian four-dimensional co-ordinate system chosen arbitrarily. That which gives the "mollusc" a certain comprehensibility as compared with the Gauss co-ordinate system is the (really unjustified) formal retention of the separate existence of the space co-ordinates as opposed to the time co-ordinate. Every point on the mollusc is treated as a space-point, and every material point which is at rest relatively to it as at rest, so long as the mollusc is considered as reference-body. The general principle of relativity requires that all these molluscs can be used as reference-bodies with equal right and equal success in the formulation of the general laws of nature; the laws themselves must be quite independent of the choice of mollusc.

The great power possessed by the general principle of relativity lies in the comprehensive limitation which is imposed on the laws of nature in consequence of what we have seen above.

29. THE SOLUTION OF THE PROBLEM OF GRAVITATION ON THE BASIS OF THE GENERAL PRINCIPLE OF RELATIVITY

If the reader has followed all our previous considerations, he will have no further difficulty in understanding the methods leading to the solution of the problem of gravitation.

We start off on a consideration of a Galileian domain, i.e. a domain in which there is no gravitational field relative to the Galileian reference-body K. The behaviour of measuring-rods and clocks with reference to K is known from the special theory of relativity, likewise the behaviour of "isolated" material points; the latter move uniformly and in straight lines.

Now let us refer this domain to a random Gauss coordinate system or to a "mollusc" as reference-body K'. Then with respect to K' there is a gravitational field G (of a particular kind). We learn the behaviour of measuring-rods and clocks and also of freely-moving material points with reference to K' simply by mathematical transformation. We interpret this behaviour as the behaviour of measuring-rods, clocks and material points under the influence of the gravitational field G. Hereupon we introduce a hypothesis: that the influence of the gravitational field on measuring-rods, clocks and freely-moving material points continues to take place according to the same laws, even in the case where the prevailing gravitational field is not derivable from the Galileian special case, simply by means of a transformation of co-ordinates.

The next step is to investigate the space-time behaviour of the gravitational field G, which was derived from the Galileian special case simply by transformation of the coordinates. This behaviour is formulated in a law, which is always valid, no matter how the reference-body (mollusc) used in the description may be chosen.

This law is not yet the general law of the gravitational field, since the gravitational field under consideration is of a special kind. In order to find out the general law-of-field of gravitation we still require to obtain a generalisation of the law as found above. This can be obtained without caprice, however, by taking into consideration the following demands:

(a) The required generalisation must likewise satisfy the general postulate of relativity.

(b) If there is any matter in the domain under consideration, only its inertial mass, and thus according to Section 15 only its energy is of importance for its effect in exciting a field.

(c) Gravitational field and matter together must satisfy the law of the conservation of energy (and of impulse).

Finally, the general principle of relativity permits us to determine the influence of the gravitational field on the course of all those processes which take place according to known laws when a gravitational field is absent i.e. which have already been fitted into the frame of the special theory of relativity. In this connection we proceed in principle according to the method which has already been explained for measuring-rods, clocks and freely moving material points.

The theory of gravitation derived in this way from the general postulate of relativity excels not only in its beauty ; nor in removing the defect attaching to classical mechanics which was brought to light in Section 21; nor in interpreting the empirical law of the equality of inertial and gravitational mass ; but it has also already explained a result of observation in astronomy, against which classical mechanics is powerless.

If we confine the application of the theory to the case where the gravitational fields can be regarded as being weak, and in which all masses move with respect to the coordinate system with velocities which are small compared with the velocity of light, we then obtain as a first approximation the Newtonian theory. Thus the latter theory is obtained here without any particular assumption, whereas Newton had to introduce the hypothesis that the force of attraction between mutually attracting material points is inversely proportional to the square of the distance between them. If we increase the accuracy of the calculation, deviations from the theory of Newton make their appearance, practically all of which must nevertheless escape the test of observation owing to their smallness.

We must draw attention here to one of these deviations. According to Newton's theory, a planet moves round the sun in an ellipse, which would permanently maintain its position with respect to the fixed stars, if we could disregard the motion of the fixed stars themselves and the action of the other planets under consideration. Thus, if we correct the observed motion of the planets for these two influences, and if Newton's theory be strictly correct, we ought to obtain for the orbit of the planet an ellipse, which is fixed with reference to the fixed stars. This deduction, which can be tested with great accuracy, has been confirmed for all the planets save one, with the precision that is capable of being obtained by the delicacy of observation attainable at the present time. The sole exception is Mercury, the planet which lies nearest the sun. Since the time of

Leverrier, it has been known that the ellipse corresponding to the orbit of Mercury, after it has been corrected for the influences mentioned above, is not stationary with respect to the fixed stars, but that it rotates exceedingly slowly in the plane of the orbit and in the sense of the orbital motion. The value obtained for this rotary movement of the orbital ellipse was 43 seconds of arc per century, an amount ensured to be correct to within a few seconds of arc. This effect can be explained by means of classical mechanics only on the assumption of hypotheses which have little probability, and which were devised solely for this purpose.

On the basis of the general theory of relativity, it is found that the ellipse of every planet round the sun must necessarily rotate in the manner indicated above ; that for all the planets, with the exception of Mercury, this rotation is too small to be detected with the delicacy of observation possible at the present time ; but that in the case of Mercury it must amount to 43 seconds of arc per century, a result which is strictly in agreement with observation.

Apart from this one, it has hitherto been possible to make only two deductions from the theory which admit of being tested by observation, to wit, the curvature of light rays by the gravitational field of the sun,[22] and a displacement of the spectral lines of light reaching us from large stars, as compared with the corresponding lines for light produced in an analogous manner terrestrially (i.e. by the same kind of atom).[23] These two deductions from the theory have both been confirmed.

Notes

22) First observed by Eddington and others in 1919. (Cf. Appendix III, pp. 126-129).

23) Established by Adams in 1924. (Cf. p. 132)

PART III : CONSIDERATIONS ON THE UNIVERSE AS A WHOLE

30. COSMOLOGICAL DIFFICULTIES OF NEWTON'S THEORY

Part from the difficulty discussed in Section 21, there is a second fundamental difficulty attending classical celestial mechanics, which, to the best of my knowledge, was first discussed in detail by the astronomer Seeliger. If we ponder over the question as to how the universe, considered as a whole, is to be regarded, the first answer that suggests itself to us is surely this: As regards space (and time) the universe is infinite. There are stars everywhere, so that the density of matter, although very variable in detail, is nevertheless on the average everywhere the same. In other words: However far we might travel through space, we should find everywhere an attenuated swarm of fixed stars of approximately the same kind and density.

This view is not in harmony with the theory of Newton. The latter theory rather requires that the universe should have a kind of centre in which the density of the stars is a maximum, and that as we proceed outwards from this centre the group-density of the stars should diminish, until finally, at great distances, it is succeeded by an infinite region of emptiness. The stellar universe ought to be a finite island in the infinite ocean of space.[24]

This conception is in itself not very satisfactory. It is still less satisfactory because it leads to the result that the light emitted by the stars and also individual stars of the stellar system are perpetually passing out into infinite space, never to return, and without ever again coming into interaction with other objects of nature. Such a finite material universe would be destined to become gradually but systematically impoverished.

In order to escape this dilemma, Seeliger suggested a modification of Newton's law, in which he assumes that for great distances the force of attraction between two masses diminishes more rapidly than would result from the inverse square law. In this way it is possible for the mean density of matter to be constant everywhere, even to infinity, without infinitely large gravitational fields being produced. We thus free ourselves from the distasteful conception that the material universe ought to possess something of the nature of a centre. Of course we purchase our emancipation from the fundamental difficulties mentioned, at the cost of a modification and complication of Newton's law which has neither empirical nor theoretical foundation. We can imagine innumerable laws which would serve the same purpose, without our being able to state a reason why one of them is to be preferred to the others ; for any one of these laws would be founded just as little on more general theoretical principles as is the law of Newton.

Notes

24) Proof--According to the theory of Newton, the number of "lines of force" which come from infinity and terminate in a mass m is proportional to the mass m. If, on the average, the Mass density ρ_0 is constant throughout the universe, then a sphere of volume V will enclose the average man $\rho_0 V$. Thus the number of lines of force passing through the surface F of the sphere into its interior is proportional to $\rho_0 V$. For unit area of the surface of the sphere the number of lines of force which enters the sphere is thus proportional to $\rho_0 V/F$ or to $\rho_0 R$. Hence the intensity of the field at the surface would ultimately become infinite with increasing radius R of the sphere, which is impossible.

31. THE POSSIBILITY OF A "FINITE" AND YET "UNBOUNDED" UNIVERSE

But speculations on the structure of the universe also move in quite another direction. The development of non-Euclidean geometry led to the recognition of the fact, that we can cast doubt on the infiniteness of our space without coming into conflict with the laws of thought or with experience (Riemann, Helmholtz). These questions have already been treated in detail and with unsurpassable lucidity by Helmholtz and Poincaré, whereas I can only touch on them briefly here.

In the first place, we imagine an existence in two dimensional space. Flat beings with flat implements, and in particular flat rigid measuring-rods, are free to move in a plane. For them nothing exists outside of this plane: that which they observe to happen to themselves and to their flat "things" is the all-inclusive reality of their plane. In particular, the constructions of plane Euclidean geometry can be carried out by means of the rods e.g. the lattice construction, considered in Section 24. In contrast to ours, the universe of these beings is two-dimensional; but, like ours, it extends to infinity. In their universe there is room for an infinite number of identical squares made up of rods, i.e. its volume (surface) is infinite. If these beings say their universe is "plane," there is sense in the statement, because they mean that they can perform the constructions of plane Euclidean geometry with their rods. In this connection the individual rods always represent the same distance, independently of their position.

Let us consider now a second two-dimensional existence, but this time on a spherical surface instead of on a plane. The flat beings with their measuring-rods and other objects fit exactly on this surface and they are unable to leave it. Their whole universe of observation extends exclusively over the surface of the sphere. Are these beings able to regard the geometry of their universe as being plane geometry and their rods withal as the realisation of "distance" ? They cannot do this. For if they attempt to realise a straight line, they will obtain a curve, which we "three-dimensional beings" designate as a great circle, i.e. a self-contained line of definite finite length, which can be measured up by means of a measuring-rod. Similarly, this universe has a finite area that can be compared with the area, of a square constructed with rods. The great charm resulting from this consideration lies in the recognition of the fact that the universe of these beings is finite and yet has no limits.

But the spherical-surface beings do not need to go on a world-tour in order to perceive that they are not living in a Euclidean universe. They can convince themselves of this on every part of their "world," provided they do not use too small a piece of it. Starting from a point, they draw "straight lines" (arcs of circles as judged in three dimensional space) of equal length in all directions. They will call the line joining the free ends of these lines a "circle." For a plane surface, the ratio of the circumference of a circle to its diameter, both lengths being measured with the same rod, is, according to Euclidean geometry of the plane, equal to a constant value π, which is independent of the diameter of the circle. On their spherical surface our flat beings would find for this ratio the value

$$\pi \frac{\sin\left(\frac{r}{R}\right)}{\left(\frac{r}{R}\right)}$$

i.e. a smaller value than π, the difference being the more considerable, the greater is the radius of the circle in comparison with the radius R of the "world-sphere." By means of this relation the spherical beings can determine the radius of their universe ("world "), even when only a relatively small part of their worldsphere is available for their measurements. But if this part is very small indeed, they will no longer be able to demonstrate that they are on a spherical "world" and not on a Euclidean plane, for a small part of a spherical surface differs only slightly from a piece of a plane of the same size.

Thus if the spherical surface beings are living on a planet of which the solar system occupies only a negligibly small part of the spherical universe, they have no means of determining whether they are living in a finite or in an infinite universe, because the "piece of universe" to which they have access is in both cases practically plane, or Euclidean. It follows directly from this discussion, that for our sphere-beings the circumference of a circle first increases with the radius until the "circumference of the universe" is reached, and that it thenceforward gradually decreases to zero for still further increasing values of the radius. During this process the area of the circle continues to increase more and more, until finally it becomes equal to the total area of the whole "world-sphere."

Perhaps the reader will wonder why we have placed our "beings" on a sphere rather than on another closed surface. But this choice has its justification in the fact that, of all closed surfaces, the sphere is unique in possessing the property that all points on it are equivalent. I admit that the ratio of the circumference c of a circle to its radius r depends on r, but for a given value of r it is the same for all points of the "worldsphere"; in other words, the

"world-sphere" is a "surface of constant curvature."

To this two-dimensional sphere-universe there is a three-dimensional analogy, namely, the three-dimensional spherical space which was discovered by Riemann. its points are likewise all equivalent. It possesses a finite volume, which is determined by its "radius" ($2\pi^2R^3$). Is it possible to imagine a spherical space? To imagine a space means nothing else than that we imagine an epitome of our "space" experience, i.e. of experience that we can have in the movement of "rigid" bodies. In this sense we can imagine a spherical space.

Suppose we draw lines or stretch strings in all directions from a point, and mark off from each of these the distance r with a measuring-rod. All the free end-points of these lengths lie on a spherical surface. We can specially measure up the area (F) of this surface by means of a square made up of measuring-rods. If the universe is Euclidean, then $F = 4\pi R^2$; if it is spherical, then F is always less than $4\pi R^2$. With increasing values of r, F increases from zero up to a maximum value which is determined by the "world-radius," but for still further increasing values of r, the area gradually diminishes to zero. At first, the straight lines which radiate from the starting point diverge farther and farther from one another, but later they approach each other, and finally they run together again at a "counter-point" to the starting point. Under such conditions they have traversed the whole spherical space. It is easily seen that the three-dimensional spherical space is quite analogous to the two-dimensional spherical surface. It is finite (i.e. of finite volume), and has no bounds.

It may be mentioned that there is yet another kind of curved space: "elliptical space." It can be regarded as a curved space in which the two "counter-points" are identical (indistinguishable from each other). An elliptical universe can thus be considered to some extent as a curved universe possessing central symmetry.

It follows from what has been said, that closed spaces without limits are conceivable. From amongst these, the spherical space (and the elliptical) excels in its simplicity, since all points on it are equivalent. As a result of this discussion, a most interesting question arises for astronomers and physicists, and that is whether the universe in which we live is infinite, or whether it is finite in the manner of the spherical universe. Our experience is far from being sufficient to enable us to answer this question. But the general theory of relativity permits of our answering it with a moderate degree of certainty, and in this connection the difficulty mentioned in Section 30 finds its solution.

32. THE STRUCTURE OF SPACE ACCORDING TO THE GENERAL THEORY OF RELATIVITY

According to the general theory of relativity, the geometrical properties of space are not independent, but they are determined by matter. Thus we can draw conclusions about the geometrical structure of the universe only if we base our considerations on the state of the matter as being something that is known. We know from experience that, for a suitably chosen co-ordinate system, the velocities of the stars are small as compared with the velocity of transmission of light. We can thus as a rough approximation arrive at a conclusion as to the nature of the universe as a whole, if we treat the matter as being at rest.

We already know from our previous discussion that the behaviour of measuring-rods and clocks is influenced by gravitational fields, i.e. by the distribution of matter. This in itself is sufficient to exclude the possibility of the exact validity of Euclidean geometry in our universe. But it is conceivable that our universe differs only slightly from a Euclidean one, and this notion seems all the more probable, since calculations show that the metrics of surrounding space is influenced only to an exceedingly small extent by masses even of the magnitude of our sun. We might imagine that, as regards geometry, our universe behaves analogously to a surface which is irregularly curved in its individual parts, but which nowhere departs appreciably from a plane: something like the rippled surface of a lake. Such a universe might fittingly be called a quasi-Euclidean universe. As regards its space it would be infinite. But calculation shows that in a quasi-Euclidean universe the average density of matter would necessarily be nil. Thus such a universe could not be inhabited by matter everywhere ; it would present to us that unsatisfactory picture which we portrayed in Section 30.

If we are to have in the universe an average density of matter which differs from zero, however small may be that difference, then the universe cannot be quasi-Euclidean. On the contrary, the results of calculation indicate that if matter be distributed uniformly, the universe would necessarily be spherical (or elliptical). Since in reality the detailed distribution of matter is not uniform, the real universe will deviate in individual parts from the spherical, i.e. the universe will be quasi-spherical. But it will be necessarily finite. In fact, the theory supplies us with a simple connection [25] between the space-expanse of the universe and the average density of matter in it.

Notes

25) For the radius R of the universe we obtain the equation

$$R^2 = \frac{2}{\kappa \rho}$$

The use of the C.G.S. system in this equation gives $2/k = 1.08 \times 10^{27}$; ρ is the average density of the matter and k is a constant connected with the Newtonian constant of gravitation.

APPENDIX I

SIMPLE DERIVATION OF THE LORENTZ TRANSFORMATION
(SUPPLEMENTARY TO SECTION 11)

For the relative orientation of the co-ordinate systems indicated in Fig. 2, the x-axes of both systems permanently coincide. In the present case we can divide the problem into parts by considering first only events which are localised on the x-axis. Any such event is represented with respect to the co-ordinate system K by the abscissa x and the time t, and with respect to the system K' by the abscissa x' and the time t'. We require to find x' and t' when x and t are given.

A light-signal, which is proceeding along the positive axis of x, is transmitted according to the equation

$$x = ct$$

or

$$x - ct = 0 \quad . \quad . \quad . \quad . \quad . \quad (1).$$

Since the same light-signal has to be transmitted relative to K' with the velocity c, the propagation relative to the system K' will be represented by the analogous formula

$$x' - ct' = 0 \quad . \quad . \quad . \quad . \quad . \quad (2)$$

Those space-time points (events) which satisfy (1) must also satisfy (2). Obviously this will be the case when the relation

$$(x' - ct') = \lambda(x - ct) \quad . \quad . \quad . \quad (3).$$

is fulfilled in general, where λ indicates a constant ; for, according to (3), the disappearance of (x - ct) involves the disappearance of (x' - ct').

If we apply quite similar considerations to light rays which are being transmitted along the negative x-axis, we obtain the condition

$$(x' + ct') = \mu(x + ct) \quad . \quad . \quad . \quad (4).$$

By adding (or subtracting) equations (3) and (4), and introducing for convenience the constants a and b in place of the constants λ and μ, where

$$a = \frac{\lambda + \mu}{2}$$

and

$$b = \frac{\lambda - \mu}{2}$$

we obtain the equations

$$\left. \begin{array}{l} x^1 = ax - bct \\ ct^1 = act - bx \end{array} \right\} \quad . \quad . \quad . \quad (5).$$

We should thus have the solution of our problem, if the constants a and b were known. These result from the following discussion.

For the origin of K' we have permanently x' = 0, and hence according to the first of the equations (5)

$$x = \frac{bc}{a} t$$

If we call v the velocity with which the origin of K' is moving relative to K, we then have

$$v = \frac{bc}{a} \qquad (6).$$

The same value v can be obtained from equations (5), if we calculate the velocity of another point of K' relative to K, or the velocity (directed towards the negative x-axis) of a point of K with respect to K'. In short, we can designate v as the relative velocity of the two systems.

Furthermore, the principle of relativity teaches us that, as judged from K, the length of a unit measuring-rod which is at rest with reference to K' must be exactly the same as the length, as judged from K', of a unit measuring-rod which is at rest relative to K. In order to see how the points of the x-axis appear as viewed from K, we only require to take a "snapshot" of K' from K; this means that we have to insert a particular value of t (time of K), e.g. t = 0. For this value of t we then obtain from the first of the equations (5)

$$x' = ax$$

Two points of the x'-axis which are separated by the distance $\Delta x' = 1$ when measured in the K' system are thus separated in our instantaneous photograph by the distance

$$\Delta x = \frac{1}{a} \qquad . \qquad . \qquad . \qquad (7).$$

But if the snapshot be taken from K'(t' = 0), and if we eliminate t from the equations (5), taking into account the expression (6), we obtain

$$x' = a\left(1 - \frac{v^2}{c^2}\right)x$$

From this we conclude that two points on the x-axis separated by the distance 1 (relative to K) will be represented on our snapshot by the distance

$$\Delta x' = a\left(1 - \frac{v^2}{c^2}\right) \qquad . \qquad . \qquad . \qquad (7a).$$

But from what has been said, the two snapshots must be identical; hence Δx in (7) must be equal to $\Delta x'$ in (7a), so that we obtain

$$a = \frac{1}{\sqrt{1 - \frac{v^2}{c^2}}} \qquad . \qquad . \qquad . \qquad (7b).$$

The equations (6) and (7b) determine the constants a and b. By inserting the values of these constants in (5), we obtain the first and the fourth of the equations given in Section 11.

$$\left.\begin{array}{l} x' = \dfrac{x - vt}{\sqrt{1 - \dfrac{v^2}{c^2}}} \\[20pt] t' = \dfrac{t - \dfrac{v}{c^2}x}{\sqrt{1 - \dfrac{v^2}{c^2}}} \end{array}\right\} \qquad . \qquad . \qquad . \qquad (8).$$

Thus we have obtained the Lorentz transformation for events on the x-axis. It satisfies the condition

$$x'^2 - c^2t'^2 = x^2 - c^2t^2 \quad . \quad . \quad . \quad (8a).$$

The extension of this result, to include events which take place outside the x-axis, is obtained by retaining equations (8) and supplementing them by the relations

$$\left.\begin{array}{l} y' = y \\ z' = z \end{array}\right\} \quad . \quad . \quad . \quad (9).$$

In this way we satisfy the postulate of the constancy of the velocity of light in vacuo for rays of light of arbitrary direction, both for the system K and for the system K'. This may be shown in the following manner.

We suppose a light-signal sent out from the origin of K at the time t= 0. It will be propagated according to the equation

$$r = \sqrt{x^2 + y^2 + z^2} = ct$$

or, if we square this equation, according to the equation

$$x^2 + y^2 + z^2 - c^2t^2 = 0 \quad . \quad . \quad . \quad (10).$$

It is required by the law of propagation of light, in conjunction with the postulate of relativity, that the transmission of the signal in question should take place--as judged from K'--in accordance with the corresponding formula

$$r' = ct'$$

or,
$$x'^2 + y'^2 + z'^2 - c^2t'^2 = 0 \quad . \quad . \quad . \quad . \quad . \quad (10a).$$

In order that equation (10a) may be a consequence of equation (10), we must have

$$x'^2 + y'^2 + z'^2 - c^2t'^2 = s\,(x^2 + y^2 + z^2 - c^2t^2) \quad (11).$$

Since equation (8a) must hold for points on the x-axis, we thus have s = 1. It is easily seen that the Lorentz transformation really satisfies equation (11) for s = 1; for (11) is a consequence of (8a) and (9), and hence also of (8) and (9). We have thus derived the Lorentz transformation.

The Lorentz transformation represented by (8) and (9) still requires to be generalised. Obviously it is immaterial whether the axes of K' be chosen so that they are spatially parallel to those of K. It is also not essential that the velocity of translation of K' with respect to K should be in the direction of the x-axis. A simple consideration shows that we are able to construct the Lorentz transformation in this general sense from two kinds of transformations, viz. from Lorentz transformations in the special sense and from purely spatial transformations. which corresponds to the replacement of the rectangular co-ordinate system by a new system with its axes pointing in other directions.

Mathematically, we can characterise the generalised Lorentz transformation thus :

It expresses x', y', x', t', in terms of linear homogeneous functions of x, y, x, t, of such a kind that the relation

$$x'^2 + y'^2 + z'^2 - c^2t'^2 = x^2 + y^2 + z^2 - c^2t^2 \qquad (11a).$$

is satisficd identically. That is to say: If we substitute their expressions in x, y, x, t, in place of x', y', x', t', on the left-hand side, then the left-hand side of (11a) agrees with the right-hand side.

APPENDIX II

MINKOWSKI'S FOUR-DIMENSIONAL SPACE ("WORLD")
(SUPPLEMENTARY TO SECTION 17)

We can characterise the Lorentz transformation still more simply if we introduce the imaginary eq. 25 in place of t, as time-variable. If, in accordance with this, we insert

$$x_1 = x$$
$$x_2 = y$$
$$x_3 = z$$
$$x_4 = \sqrt{-1} \cdot ct \dots\dots\text{eq. 25}$$

and similarly for the accented system K', then the condition which is identically satisfied by the transformation can be expressed thus :

$$x_1'^2 + x_2'^2 + x_3'^2 + x_4'^2 = x_1^2 + x_2^2 + x_3^2 + x_4^2 \qquad (12).$$

That is, by the afore-mentioned choice of "coordinates," (11a) [see the end of Appendix II] is transformed into this equation.

We see from (12) that the imaginary time co-ordinate x_4, enters into the condition of transformation in exactly the same way as the space co-ordinates x_1, x_2, x_3. It is due to this fact that, according to the theory of relativity, the "time" x_4, enters into natural laws in the same form as the space co ordinates x_1, x_2, x_3.

A four-dimensional continuum described by the "co-ordinates" x_1, x_2, x_3, x_4, was called "world" by Minkowski, who also termed a point-event a "world-point." From a "happening" in three-dimensional space, physics becomes, as it were, an "existence" in the four-dimensional "world."

This four-dimensional "world" bears a close similarity to the three-dimensional "space" of (Euclidean) analytical geometry. If we introduce into the latter a new Cartesian co-ordinate system (x'_1, x'_2, x'_3) with the same origin, then x'_1, x'_2, x'_3, are linear homogeneous functions of x_1, x_2, x_3 which identically satisfy the equation

$$x'^2_1 + x'^2_2 + x'^2_3 = x^2_1 + x^2_2 + x^2_3$$

The analogy with (12) is a complete one. We can regard Minkowski's "world" in a formal manner as a four-dimensional Euclidean space (with an imaginary time coordinate) ; the Lorentz transformation corresponds
to a "rotation" of the co-ordinate system in the four-dimensional "world."

APPENDIX III

THE EXPERIMENTAL CONFIRMATION OF THE GENERAL THEORY OF RELATIVITY

From a systematic theoretical point of view, we may imagine the process of evolution of an empirical science to be a continuous process of induction. Theories are evolved and are expressed in short compass as statements of a large number of individual observations in the form of empirical laws, from which the general laws can be ascertained by comparison. Regarded in this way, the development of a science bears some resemblance to the compilation of a classified catalogue. It is, as it were, a purely empirical enterprise.

But this point of view by no means embraces the whole of the actual process ; for it slurs over the important part played by intuition and deductive thought in the development of an exact science. As soon as a science has emerged from its initial stages, theoretical advances are no longer achieved merely by a process of arrangement. Guided by empirical data, the investigator rather develops a system of thought which, in general, is built up logically from a small number of fundamental assumptions, the so-called axioms. We call such a system of thought a theory. The theory finds the justification for its existence in the fact that it correlates a large number of single observations, and it is just here that the "truth" of the theory lies.

Corresponding to the same complex of empirical data, there may be several theories, which differ from one another to a considerable extent. But as regards the deductions from the theories which are capable of being tested, the agreement between the theories may be so complete that it becomes difficult to find any deductions in which the two theories differ from each other. As an example, a case of general interest is available in the province of biology, in the Darwinian theory of the development of species by selection in the struggle for existence, and in the theory of development which is based on the hypothesis of the hereditary transmission of acquired characters.

We have another instance of far-reaching agreement between the deductions from two theories in Newtonian mechanics on the one hand, and the general theory of relativity on the other. This agreement goes so far, that up to the present we have been able to find only a few deductions from the general theory of relativity which are capable of investigation, and to which the physics of pre-relativity days does not also lead, and this despite the profound difference in the fundamental assumptions of the two theories. In what follows, we shall again consider these important deductions, and we shall also discuss the empirical evidence appertaining to them which has hitherto been obtained.

(a) Motion of the Perihelion of Mercury

According to Newtonian mechanics and Newton's law of gravitation, a planet which is revolving round the sun would describe an ellipse round the latter, or, more correctly, round the common centre of gravity of the sun and the planet. In such a system, the sun, or the common centre of gravity, lies in one of the foci of the orbital ellipse in such a manner that, in the course of a planet-year, the distance sun-planet grows from a minimum to a maximum, and then decreases again to a minimum. If instead of Newton's law we insert a somewhat different law of attraction into the calculation, we find that, according to this new law, the motion would still take place in such a manner that the distance sun-planet exhibits periodic variations; but in this case the angle described by the line joining sun and planet during such a period (from perihelion--closest proximity to the sun--to perihelion) would differ from 360°. The line of the orbit would not then be a closed one but in the course of time it would fill up an annular part of the orbital plane, viz. between the circle of least and the circle of greatest distance of the planet from the sun.

According also to the general theory of relativity, which differs of course from the theory of Newton, a small variation from the Newton-Kepler motion of a planet in its orbit should take place, and in such away, that the angle described by the radius sun-planet between one perhelion and the next should exceed that corresponding to one complete revolution by an amount given by

$$+ \ \frac{24 \ \pi^3 a^2}{T^2 c^2 (1 - e^2)}$$

(N.B.--One complete revolution corresponds to the angle 2π in the absolute angular measure customary in physics, and the above expression given the amount by which the radius sun-planet exceeds this angle during the interval between one perihelion and the next.) In this expression a represents the major semi-axis of the ellipse, e its eccentricity, c the velocity of light, and T the period of revolution of the planet. Our result may also be stated as follows : According to the general theory of relativity, the major axis of the ellipse rotates round the sun in the same sense as the orbital motion of the planet. Theory requires that this rotation should amount to 43 seconds of arc per century for the planet Mercury, but for the other Planets of our solar system its magnitude should be so small that it would necessarily escape detection.[26]

In point of fact, astronomers have found that the theory of Newton does not suffice to calculate the observed motion of Mercury with an exactness corresponding to that of the delicacy of observation attainable at the present time. After taking account of all the disturbing influences exerted on Mercury by the remaining planets, it was found (Leverrier: 1859; and Newcomb: 1895) that an unexplained perihelial movement of the orbit of Mercury remained over, the amount of which does not differ sensibly from the above mentioned +43 seconds of arc per century. The uncertainty of the empirical result amounts to a few seconds only.

(b) Deflection of Light by a Gravitational Field

In Section 22 it has been already mentioned that according to the general theory of relativity, a ray of light will experience a curvature of its path when passing through a gravitational field, this curvature being similar to that experienced by the path of a body which is projected through a gravitational field. As a result of this theory, we should expect that a ray of light which is passing close to a heavenly body would be deviated towards the latter. For a ray of light which passes the sun at a distance of Δ sun-radii from its centre, the angle of deflection (a) should amount to

$$a = \frac{1.7 \ \text{seconds of arc}}{\Delta}$$

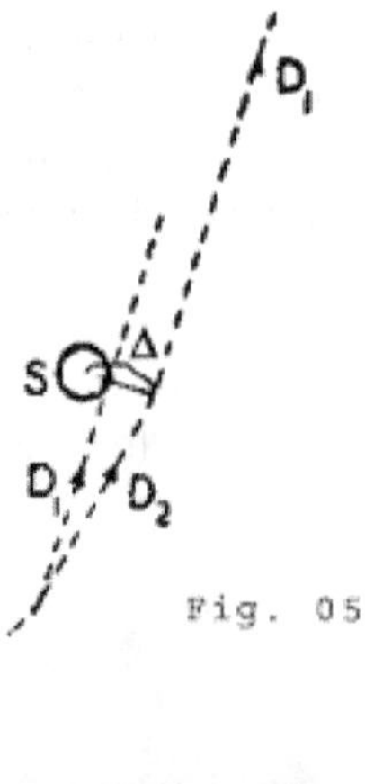

Fig. 05

It may be added that, according to the theory, half of this deflection is produced by the Newtonian field of attraction of the sun, and the other half by the geometrical modification ("curvature") of space caused by the sun.

This result admits of an experimental test by means of the photographic registration of stars during a total eclipse of the sun. The only reason why we must wait for a total eclipse is because at every other time the atmosphere is so strongly illuminated by the light from the sun that the stars situated near the sun's disc are invisible. The predicted effect can be seen clearly from the accompanying diagram. If the sun (S) were not present, a star which is practically infinitely distant would be seen in the direction D_1, as observed front the earth. But as a consequence of the deflection of light from the star by the sun, the star will be seen in the direction D_2, i.e. at a somewhat greater distance from the centre of the sun than corresponds to its real position.

In practice, the question is tested in the following way. The stars in the neighbourhood of the sun are photographed during a solar eclipse. In addition, a second photograph of the same stars is taken when the sun is situated at another position in the sky, i.e. a few months earlier or later. As compared with the standard photograph, the positions of the stars on the eclipse-photograph ought to appear displaced radially outwards (away from the centre of the sun) by an amount corresponding to the angle a.

We are indebted to the [British] Royal Society and to the Royal Astronomical Society for the investigation of this important deduction. Undaunted by the [first world] war and by difficulties of both a material and a psychological nature aroused by the war, these societies equipped two expeditions--to Sobral (Brazil), and to the island of Principe (West Africa)--and sent several of Britain's most celebrated astronomers (Eddington, Cottingham, Crommelin, Davidson), in order to obtain photographs of the solar eclipse of 29th May, 1919. The relative discrepancies to be expected between the stellar photographs obtained during the eclipse and the comparison photographs amounted to a few hundredths of a millimetre only. Thus great accuracy was necessary in making the adjustments required for the taking of the photographs, and in their subsequent measurement.

The results of the measurements confirmed the theory in a thoroughly satisfactory manner. The rectangular components of the observed and of the calculated deviations of the stars (in seconds of arc) are set forth in the following table of results :

Number of the Star.	First Co-ordinate.		Second Co-ordinate	
	Observed.	Calculated.	Observed.	Calculated.
11	−0·19	−0·22	+0·16	+0·02
5	+0·29	+0·31	−0·46	−0·43
4	+0·11	+0·10	+0·83	+0·74
3	+0·20	+0·12	+1·00	+0·87
6	+0·10	+0·04	+0·57	+0·40
10	−0·08	+0·09	+0·35	+0·32
2	+0·95	+0·85	−0·27	−0·09

(c) Displacement of Spectral Lines Towards the Red

In Section 23 it has been shown that in a system K' which is in rotation with regard to a Galileian system K, clocks of identical construction, and which are considered at rest with respect to the rotating reference-body, go at rates which are dependent on the positions of the clocks. We shall now examine this dependence quantitatively. A clock, which is situated at a distance r from the centre of the disc, has a velocity relative to K which is given by

$$V = wr$$

where w represents the angular velocity of rotation of the disc K' with respect to K. If v_0, represents the number of ticks of the clock per unit time ("rate" of the clock) relative to K when the clock is at rest, then the "rate" of the clock (v) when it is moving relative to K with a velocity V, but at rest with respect to the disc, will, in accordance with Section 12, be given by

$$v = v_2 \sqrt{1 - \frac{V^2}{c^2}}$$

or with sufficient accuracy by

$$v = v_0 \left(1 - \frac{1}{2}\frac{V^2}{c^2}\right)$$

This expression may also be stated in the following form:

$$v = v_0 \left(1 - \frac{1}{c^2}\frac{w^2 r^2}{2}\right)$$

If we represent the difference of potential of the centrifugal force between the position of the clock and the centre of the disc by f, i.e. the work, considered negatively, which must be performed on the unit of mass against the centrifugal force in order to transport it from the position of the clock on the rotating disc to the centre of the disc, then we have

$$\phi = \frac{w^2 r^2}{2}$$

From this it follows that

$$v = v_0 \left(1 + \frac{\phi}{c^2}\right)$$

In the first place, we see from this expression that two clocks of identical construction will go at different rates when situated at different distances from the centre of the disc. This result is also valid from the standpoint of an observer who is rotating with the disc.

Now, as judged from the disc, the latter is in a gravitational field of potential f, hence the result we have obtained will hold quite generally for gravitational fields. Furthermore, we can regard an atom which is emitting spectral lines as a clock, so that the following statement will hold:

An atom absorbs or emits light of a frequency which is dependent on the potential of the gravitational field in which it is situated.

The frequency of an atom situated on the surface of a heavenly body will be somewhat less than the frequency of an atom of the same element which is situated in free space (or on the surface of a smaller celestial body).

Now $f = - K (M/r)$, where K is Newton's constant of gravitation, and M is the mass of the heavenly body. Thus a displacement towards the red ought to take place for spectral lines produced at the surface of stars as compared with the spectral lines of the same element produced at the surface of the earth, the amount of this displacement being

$$\frac{v_0 - v}{v_0} = \frac{K}{c^2} \frac{M}{r}$$

For the sun, the displacement towards the red predicted by theory amounts to about two millionths of the wave-length. A trustworthy calculation is not possible in the case of the stars, because in general neither the mass M nor the radius r are known.

It is an open question whether or not this effect exists, and at the present time (1920) astronomers are working with great zeal towards the solution. Owing to the smallness of the effect in the case of the sun, it is difficult to form an opinion as to its existence.

Whereas Grebe and Bachem (Bonn), as a result of their own measurements and those of Evershed and Schwarzschild on the cyanogen bands, have placed the existence of the effect almost beyond doubt, while other investigators, particularly St. John, have been led to the opposite opinion in consequence of their measurements.

Mean displacements of lines towards the less refrangible end of the spectrum are certainly revealed by statistical investigations of the fixed stars ; but up to the present the examination of the available data does not allow of any definite decision being arrived at, as to whether or not these displacements are to be referred in reality to the effect of gravitation. The results of observation have been collected together, and discussed in detail from the standpoint of the question which has been engaging our attention here, in a paper by E. Freundlich entitled "Zur Prüfung der allgemeinen Relativitäts-Theorie" (Die Naturwissenschaften, 1919, No. 35, p. 520: Julius Springer, Berlin).

At all events, a definite decision will be reached during the next few years. If the displacement of spectral lines towards the red by the gravitational potential does not exist, then the general theory of relativity will be untenable. On the other hand, if the cause of the displacement of spectral lines be definitely traced to the gravitational potential, then the study of this displacement will furnish us with important information as to the mass of the heavenly bodies. [27)]

Notes

26) Especially since the next planet Venus has an orbit that is almost an exact circle, which makes it more difficult to locate the perihelion with precision.

27) The displacement of spectral lines towards the red end of the spectrum was definitely established by Adams in 1924, by observations on the dense companion of Sirius, for which the effect is about thirty times greater than for the Sun. R.W.L.-- translator

APPENDIX IV

THE STRUCTURE OF SPACE ACCORDING TO THE GENERAL THEORY OF RELATIVITY
(SUPPLEMENTARY TO SECTION 32)

Since the publication of the first edition of this little book, our knowledge about the structure of space in the large ("cosmological problem") has had an important development, which ought to be mentioned even in a popular presentation of the subject.

My original considerations on the subject were based on two hypotheses:

(1) There exists an average density of matter in the whole of space which is everywhere the same and different from zero.

(2) The magnitude ("radius") of space is independent of time.

Both these hypotheses proved to be consistent, according to the general theory of relativity, but only after a hypothetical term was added to the field equations, a term which was not required by the theory as such nor did it seem natural from a theoretical point of view ("cosmological term of the field equations").

Hypothesis (2) appeared unavoidable to me at the time, since I thought that one would get into bottomless speculations if one departed from it.

However, already in the 'twenties, the Russian mathematician Friedman showed that a different hypothesis was natural from a purely theoretical point of view. He realized that it was possible to preserve hypothesis (1) without introducing the less natural cosmological term into the field equations of gravitation, if one was ready to drop hypothesis (2). Namely, the original field equations admit a solution in which the "world radius" depends on time (expanding space). In that sense one can say, according to Friedman, that the theory demands an expansion of space.

A few years later Hubble showed, by a special investigation of the extra-galactic nebulae ("milky ways"), that the spectral lines emitted showed a red shift which increased regularly with the distance of the nebulae. This can be interpreted in regard to our present knowledge only in the sense of Doppler's principle, as an expansive motion of the system of stars in the large--as required, according to Friedman, by the field equations of gravitation. Hubble's discovery can, therefore, be considered to some extent as a confirmation of the theory.

There does arise, however, a strange difficulty. The interpretation of the galactic line-shift discovered by Hubble as an expansion (which can hardly be doubted from a theoretical point of view), leads to an origin of this expansion which lies "only" about 10^9 years ago, while physical astronomy makes it appear likely that the development of individual stars and systems of stars takes considerably longer. It is in no way known how this incongruity is to be overcome.

I further want to remark that the theory of expanding space, together with the empirical data of astronomy, permit no decision to be reached about the finite or infinite character of (three-dimensional) space, while the original "static" hypothesis of space yielded the closure (finiteness) of space.

K = co-ordinate system
x, y = two-dimensional co-ordinates
x, y, z = three-dimensional co-ordinates
x, y, z, t = four-dimensional co-ordinates

t = time
I = distance
v = velocity

F = force
G = gravitational field

www.ingramcontent.com/pod-product-compliance
Lightning Source LLC
Chambersburg PA
CBHW081433170526
45166CB00008B/2193